Extreme Floods and Droughts under Future Climate Scenarios

Extreme Floods and Droughts under Future Climate Scenarios

Special Issue Editors

Momcilo Markus
Ximing Cai
Ryan Sriver

MDPI • Basel • Beijing • Wuhan • Barcelona • Belgrade

MDPI

Special Issue Editors
Momcilo Markus
University of Illinois
USA

Ximing Cai
University of Illinois
USA

Ryan Sriver
University of Illinois
USA

Editorial Office
MDPI
St. Alban-Anlage 66
4052 Basel, Switzerland

This is a reprint of articles from the Special Issue published online in the open access journal *Water* (ISSN 2073-4441) from 2018 to 2019 (available at: https://www.mdpi.com/journal/water/special_issues/floods_droughts).

For citation purposes, cite each article independently as indicated on the article page online and as indicated below:

LastName, A.A.; LastName, B.B.; LastName, C.C. Article Title. *Journal Name* **Year**, *Article Number*, Page Range.

ISBN 978-3-03921-898-1 (Pbk)
ISBN 978-3-03921-899-8 (PDF)

Contents

About the Special Issue Editors

Momcilo Markus is a hydrologist at the Illinois State Water Survey, as well as a Research Associate Professor of the Department of Agricultural and Biological Engineering, and the Department of Natural Resources and Environmental Sciences at the University of Illinois at Urbana-Champaign. His current work includes statistical hydrology, data analytics and hydroclimatology.

Ximing Cai is Lovell Endowed Professor of Civil and Environmental Engineering at the University of Illinois at Urbana-Champaign. His current research includes drought management, using forecasts to control extreme weather events, water resources system operation, and interdependent infrastructure system analysis.

Ryan Sriver is an Associate Professor in the Department of Atmospheric Sciences at the University of Illinois of Urbana-Champaign. His research interests include climate dynamics and variability, Earth system modeling, climate and weather extremes, and uncertainty quantification.

water

MDPI

Editorial

Extreme Floods and Droughts under Future Climate Scenarios

Momcilo Markus [1,*], **Ximing Cai** [2] **and Ryan Sriver** [3]

[1]   Prairie Research Institute, University of Illinois, Urbana, IL 61801, USA
[2]   Department of Civil and Environmental Engineering, University of Illinois, Urbana, IL 61801, USA
[3]   Department of Atmospheric Sciences, University of Illinois, Urbana, IL 61801, USA
*   Correspondence: mmarkus@illinois.edu; Tel.: +1-217-333-0237

Received: 12 August 2019; Accepted: 12 August 2019; Published: 19 August 2019

Abstract: Climate projections indicate that in many regions of the world the risk of increased flooding or more severe droughts will be higher in the future. To account for these trends, hydrologists search for the best planning and management measures in an increasingly complex and uncertain environment. The collection of manuscripts in this Special Issue quantifies the changes in projected hydroclimatic extremes and their impacts using a suite of innovative approaches applied to regions in North America, Asia, and Europe. To reduce the uncertainty and warrant the applicability of the research on projections of future floods and droughts, their continued development and testing using newly acquired observational data are critical.

Keywords: climate change; climate projections; extreme rainfall; floods; droughts

1. Introduction

The purpose of this Special Issue is to highlight several innovative manuscripts with a focus on projected hydroclimatic extremes and their impacts. Special attention was given to manuscripts quantifying projected changes in extreme precipitation and droughts, floods, and water quality parameters.

In many geographic regions, the effects of climate change on hydrologic systems are projected to be significant. Determining those regions and the direction and magnitude of changes is critical for long-term water management and planning decisions. Additionally, these decisions require the accuracy of projections that are often based on limited data and numerical models with large uncertainties. Complexities in the hydroclimatic processes, however, make estimating aggregate uncertainties surrounding projections of historical and future climate variability and change difficult and challenging. Characterizing changes in extreme events, such as temperature and precipitation extremes, floods, and droughts, is particularly difficult on account of small sample sizes and relatively short observational records. In addition, variability in large-scale dynamics and teleconnections may influence regional temperature and precipitation extremes across different climate model ensembles [1,2] because of differences in resolution, model structure, and the representation of teleconnections between ocean temperature, polar sea ice, and mid-latitude jet stream variability [3]. Even with these uncertainties, we observe robust changes in key climate variables related to floods and droughts, including increases in extreme warm temperatures globally and regional changes in the distributions of heavy precipitation events [4,5]. Changes in extremes over time are not solely due to shifts in the means of the distributions; they can also be associated with changes in shape and skewness related to the tails/extremes. Changes in sequencing and compound extreme events (e.g., drought combined with a heatwave) pose additional risks that are not well quantified in observations and models [6]. Special emphasis needs to be given to increasing modeling accuracy and proper determination of the confidence in the results. Continuous monitoring of climatic variables, the refinement of climate models, and the development of statistical/stochastic methods will be key components in more accurately determining the effects of climate change on hydrologic and climatic extremes.

2. Summary of the Papers in the Special Issue

The papers in this Special Issue are well-balanced in terms of their focus, ranging from heavy precipitation and floods to droughts and water quality. The papers also cover a wide range of climates and geographic regions, including Canada, the USA, Poland, and China.

Two papers [7,8] address heavy precipitation. In [7], the authors evaluate the applicability of the spatial analog as an alternative approach in extreme precipitation analysis using data from the central USA. They also highlight the existence of large uncertainties in studying climate projections and the need to minimize them in the future. In another paper [8], a comparative analysis of four downscaled data sets shows that a newly designed, novel statistical downscaling data set favorably compared with other commonly used data sets when applied to extreme rainfall analysis. This research also determined the magnitude of the projected increase in heavy precipitation in the Northeastern USA. For example, they determined that the event with a recurrence interval of 100 years today will have a 19- to 25-year recurrence interval in the late 21st century, depending on the climate scenarios.

A majority of the Special Issue authors presented their research dealing with projected streamflow extremes in various rivers around the world [9–12]. The projected streamflow characteristics, including frequency and timing of future flooding in Canada, were calculated [9]. This paper determined the changes for all regions in the country and provided a comprehensive assessment of the uncertainties in their estimates. In a companion paper [10], the authors discuss the consequences of these changes on flow regulation infrastructures (FRI) in highly populated/urban areas. They found that flood management guidelines for some FRIs would have to be reassessed to make them resilient to increased flooding in the future. In a paper with a similar scope [11], CMIP5-based climate modeling data were applied to a hydrologic model to produce future streamflow ensembles for a watershed in Idaho, USA. These ensembles were a broader envelope of the historic data, but the results generally indicated that flash flooding will increase in the future. The effects of errors in riverine flow projections on the return period of projected annual maximum inundation areas were presented in [12]. This research was based on a distributed flow routing model, MIKE11, and its lumped-parameter emulator to illustrate this uncertainty using a watershed in Southern Poland.

Papers [13,14] reported research on droughts. Future droughts under a range of climate scenarios were examined in [13]. This research identified the season, direction, and magnitude of changes in the length of dry spells for different parts of the Southeast United States. The effects of droughts on both low flows and floods were reported in [14]. This paper presented a novel method to create a super-ensemble of future climate series, which was then used to derive flooding and drought indices. The results indicated a marginal increase in low flows and a small decrease in high flows on the South Nation Watershed located in Ontario, Canada.

The drought-flood abrupt alteration (DFAA) has been studied as an extreme hydrological phenomenon, which is particularly important in riverine nutrient loading [15]. The results show projected changes in DFAA in the Hetao area, China, highlighting the often-neglected, but critically important effects of climate variability and change on riverine water quality, and implying combined drainage and irrigation measures during crop growth seasons.

3. Reflection of Future Research

Due to climatic changes, the risk of hydroclimatic extremes will be different in the future in many geographic regions of the world. To better prepare for the changes, it is important to account for impacts at the regional-to-local scales. Climate projections are typically based upon observational data analysis, climate monitoring, and relatively coarse climate models. These products can be combined using statistical and dynamical downscaling techniques to provide climate information at finer spatial and temporal scales. The accuracy of model-based climate projects is typically assessed through comparisons with historical data. It is generally assumed that the historical accuracy of climate models is an indication of future performance, but more research is required to justify this assumption. Consequently, models with higher historical accuracy often receive higher weights in climate studies, but more studies are needed to determine whether weighted or equal weights should be used, and, if so, how the weights should be defined, and what would be the benefit of weights. Climate model products are also inherently uncertain, since the models represent simplified versions of the real world. These uncertainties can be due to coarse model resolution, a lack of physics or necessary complexity, forcing uncertainties, structural model differences, and different representations of natural variability (e.g., internal model variability).

Recent climate downscaling techniques are capable of producing increasingly finer spatial and temporal resolutions for climatic variables. These products are typically finer-scale resolution than global models. For example, the NASA Earth Exchange (NEX) dynamically downscaled products, based on adjusting statistical moments in temperature and precipitation outputs from CMIP5 models, are formatted with 25 km horizontal grid resolution, whereas the global models are typically on the order of 1 to 2 degrees. These downscaled products can be used to reduce model biases based on the agreement with historical records, but it is not clear how the corrections may influence the uncertainty in future projections, particularly if the statistical moments are changing with time. These challenges are exacerbated when considering extreme events, given the relatively short observational record, inadequacies of the models to represent extremes, and uncertainties in how tail parameters are changing.

Extreme weather events can also be jointly affected by both the influence of global warming and internal variabilities, such as the El Niño Southern–Oscillation (ENSO), Atlantic Multidecadal Oscillation (AMO), and Pacific Decadal Oscillation (PDO). In many cases, the internal variability appeared as the major driving force [16].

Generally, two main approaches are used in studying the effects of climate change on the frequency of floods and droughts. One approach assumes two or more quasi-stationary time periods and treats each of them as stationary. The other approach expresses statistics (e.g., frequency distribution parameters) as a function of time. Both approaches have advantages and limitations. Site-specific non-stationary frequency analysis is prone to having large uncertainties. The distribution parameter estimates, e.g., the shape parameter of a frequently used generalized extreme value distribution, can be sensitive to large observations, resulting in very uncertain quantile estimates. This problem could be alleviated by limiting the parameter ranges, or by developing a regional analysis instead of point estimates. For any quasi-stationary approach, assuming stationarity for longer time periods is not realistic in a changing environment. On the other hand, using shorter time periods for statistical analysis [8,17,18] creates a problem with estimations of large recurrent events (e.g., 100-year event based on 20 years). Thus, the stationarity assumptions need to be evaluated in each case (e.g., the evaluation of the stationarity assumption for meteorological drought estimation in the contiguous US [19]).

For rainfall and runoff frequency analysis, the annual maximum series (AMS) and partial duration series (PDS) methods for frequency estimates are generally used. Typically, the results indicate that the AMS with Langbein's adjustment [20] would produce very similar results to those based on PDS, but this does not have to be true for projected extremes. An additional reason for using PDS instead of AMS could be the fact that the increases in precipitation frequency and magnitude are not correlated [21]. It would be very useful to validate the findings of the paper as new data become available.

Estimating confidence limits is possibly the most challenging task in the analysis of weather/climate extremes related to floods and drought. Many existing methods incorporate calculations of confidence intervals (CIs), but they tend to be overconfident and underestimate the true limits. For example, the L-moments approach creates CIs based on Monte Carlo resampling of statistical distribution parameters, but it ignores the uncertainty based on the distribution selection, data observations, their spatial/temporal aggregation, etc.

The propagation of uncertainties from the global climate to regional extremes also needs more attention, as well as the cross-auto-correlation structure of residuals, using hybrid approaches capable of synthesizing observations and models [22,23]. Coupled feedbacks and interactions between natural and human systems can also influence the occurrence and severity of extreme climate/weather events. New multidisciplinary research and methodologies in areas related to extremes are broadly applicable to water planning and management, which can ultimately lead to improved optimum water system design and control in a changing environment.

Conflicts of Interest: The authors declare no conflict of interest.

References

1. Sriver, R.L.; Forest, C.E.; Keller, K. Effects of Initial Conditions Uncertainty on Regional Climate Variability: An Analysis Using a Low-Resolution CESM Ensemble. *Geophys. Res. Lett.* **2015**, *42*, 5468–5476. [CrossRef]
2. Hogan, E.E.; Nicholas, R.; Keller, K.; Eilts, S.; Sriver, R.L. Representation of US Warm Temperature Extremes in Global Climate Model Ensembles. *J. Clim.* **2019**, *32*, 2591–2603. [CrossRef]
3. Francis, J.A. Why Are Arctic Linkages to Extreme Weather Still up in the Air? *Bull. Am. Meteorol. Soc.* **2017**, *98*, 2551–2557. [CrossRef]
4. IPCC. Summary for policymakers. In *Managing the Risks of Extreme Events and Disasters to Advance Climate Change Adaptation. A Special Report of Working Groups I and II of the Intergovernmental Panel on Climate Change*; Field, C.B., Barros, V., Stocker, T.F., Qin, D., Dokken, D.J., Ebi, K.L., Mastrandrea, M.D., Mach, K.J., Plattner, G.-K., Allen, S.K., et al., Eds.; Cambridge University Press: Cambridge, UK; New York, NY, USA, 2012; pp. 3–21.
5. Vose, R.S.; Easterling, D.R.; Kunkel, K.E.; LeGrande, A.N.; Wehner, M.F. Temperature changes in the United States. In *Climate Science Special Report: Fourth National Climate Assessment, Volume, I.*; Wuebbles, D.J., Fahey, D.W., Hibbard, K.A., Dokken, D.J., Stewart, B.C., Maycock, T.K., Eds.; U.S. Global Change Research Program: Washington, DC, USA, 2017; pp. 185–206.
6. Kopp, R.E.; Hayhoe, K.; Easterling, D.R.; Hall, T.; Horton, R.; Kunkel, K.E.; LeGrande, A.N. Potential surprises: Compound extremes and tipping elements. In *Climate Science Special Report: Fourth National Climate Assessment, Volume, I.*; Wuebbles, D.J., Fahey, D.W., Hibbard, K.A., Dokken, D.J., Stewart, B.C., Maycock, T.K., Eds.; U.S. Global Change Research Program: Washington, DC, USA, 2017; pp. 411–429.
7. Wang, A.K.; Dominguez, F.; Schmidt, A.R. Extreme Precipitation Spatial Analog: In Search of an Alternative Approach for Future Extreme Precipitation in Urban Hydrological Studies. *Water* **2019**, *11*, 1032. [CrossRef]
8. Wu, S.; Markus, M.; Lorenz, D.; Angel, J.R.; Grady, K. A Comparative Analysis of the Hindcast Accuracy of the Point Precipitation Frequency Estimates of Four Data Sets and Their Projections for the Northeastern United States. *Water* **2019**, *11*, 1279. [CrossRef]
9. Gaur, A.; Gaur, A.; Simonovic, S.P. Future Changes in Flood Hazard Across Canada Under Changing Climate. *Water* **2018**, *10*, 1441. [CrossRef]
10. Gaur, A.; Gaur, A.; Yamazaki, D.; Simonovic, S.P. Flooding Related Consequences of Climate Change on Canadian Cities and Flow Regulation Infrastructure. *Water* **2019**, *11*, 63. [CrossRef]
11. Ryu, J.; Kim, J. A Study on Climate-Driven Flood Risks in the Boise River Watershed, Idaho. *Water* **2019**, *11*, 1039. [CrossRef]
12. Doroszkiewicz, J.; Romanowicz, R.J.; Kiczko, A. An Influence of Flow Projection Errors on Flood Hazard Estimates in Future Climate Conditions. *Water* **2019**, *11*, 49. [CrossRef]
13. Keellings, D.; Engström, J. The Future of Drought in the Southeastern, U.S.: Projections from Downscaled CMIP5 Models. *Water* **2019**, *11*, 259. [CrossRef]
14. Alodah, A.; Seidou, O. Assessment of Climate Change Impacts on Extreme High and Low Flows by an Improved Bottom-up Approach. *Water* **2019**, *11*, 1236. [CrossRef]
15. Yang, Y.; Weng, B.; Bi, W.; Xu, T.; Yan, D.; Ma, J. Impacts of Future Climate Change on Drought-Flood Abrupt Alternation and Water Quality in the Hetao Area, China. *Water* **2019**, *11*, 652. [CrossRef]
16. Apurv, T.; Cai, X.; Yuan, X. Influence of Internal Variability and Global Warming on Multidecadal Changes in Regional Drought Severity Over the Continental, U.S. *J. Hydrometeorol.* **2019**, *20*, 411–429. [CrossRef]

17. Markus, M.; Angel, J.; Byard, G.; McConkey, S.; Zhang, C.; Cai, X.; Notaro, M.; Ashfaq, M. Communicating the Impacts of Projected Climate Change on Heavy Rainfall Using a Weighted Ensemble Approach. *J. Hydrol. Eng.* **2018**, *23*, 04018004. [CrossRef]
18. Markus, M.; Wuebbles, D.J.; Liang, X.-Z.; Hayhoe, K.; Kristovich, D.A.R. Diagnostic analysis of future climate scenarios applied to urban flooding in the Chicago metropolitan area. *Clim. Chang.* **2012**, *111*, 879–902. [CrossRef]
19. Apurv, T.; Cai, X. Evaluation of the Stationarity Assumption for Meteorological Drought Risk Estimation at the Multidecadal Scale in Contiguous United States. *Water Resour. Res.* **2019**, *55*, 5074–5101.
20. Langbein, W.B. Annual Floods and the Partial-Duration Flood Series. *Trans. Am. Geophys. Union* **1949**, *30*, 879–881. [CrossRef]
21. Papalexiou, A.; Montanari, A. Global and Regional Increase of Precipitation Extremes under Global Warming. *Water Resour. Res.* **2019**, *55*. [CrossRef]
22. Haugen, M.A.; Stein, M.L.; Moyer, E.J.; Sriver, R.L. Estimating Changes in Temperature Distributions in a Large Ensemble of Climate Simulations Using Quantile Regression. *J. Clim.* **2018**, *31*, 8573–8588. [CrossRef]
23. Haugen, M.A.; Stein, M.L.; Sriver, R.L.; Moyer, E.L. Future Climate Emulations Using Quantile Regressions on Large Ensembles. *Adv. Stat. Climatol. Meteorol. Oceanogr.* **2019**, *5*, 37–55. [CrossRef]

water

MDPI

Article

Future Changes in Flood Hazards across Canada under a Changing Climate

Ayushi Gaur [1], Abhishek Gaur [2,*] and Slobodan P. Simonovic [1]

[1] Facility for Intelligent Decision Support, Department of Civil and Environmental Engineering, The University of Western, London, Ontario, ON N6A 3K7, Canada; ayushigaur.evs@gmail.com (A.G.); simonovic@uwo.ca (S.P.S.)

[2] National Research Council Canada, 1200 Montreal Road, Ottawa, ON K1A 0R6, Canada

* Correspondence: Abhishek.Gaur@nrc-cnrc.gc.ca

Received: 4 September 2018; Accepted: 7 October 2018; Published: 13 October 2018

Abstract: Climate change has induced considerable changes in the dynamics of key hydro-climatic variables across Canada, including floods. In this study, runoff projections made by 21 General Climate Models (GCMs) under four Representative Concentration Pathways (RCPs) are used to generate 25 km resolution streamflow estimates across Canada for historical (1961–2005) and future (2061–2100) time-periods. These estimates are used to calculate future projected changes in flood magnitudes and timings across Canada. Results obtained indicate that flood frequencies in the northernmost regions of Canada, and south-western Ontario can be expected to increase in the future. As an example, the historical 100-year return period events in these regions are expected to become 10–60 year return period events. On the other hand, northern prairies and north-central Ontario can be expected to experience decreases in flooding frequencies in future. The historical 100-year return period flood events in these regions are expected to become 160–200 year return period events in future. Furthermore, prairies, parts of Quebec, Ontario, Nunavut, and Yukon territories can be expected to experience earlier snowmelt-driven floods in the future. The results from this study will help decision-makers to effectively manage and design municipal and civil infrastructure in Canada under a changing climate.

Keywords: climate change; Canada; flooding frequency; catchment based macroscale floodplain model; uncertainty

1. Introduction

Floods are the most frequently occurring natural hazard in Canada and around the globe [1–4]. A number of studies have been performed in different parts of the globe to establish methods for effective quantification of floods and their associated risks [5–17]. Studies have also investigated methods to quantify compound flooding i.e., that are caused by two or more events contributing to flooding example occurrence of extreme rainfall, variations in astronomical tides, storm surge, and wave action, rise in groundwater levels etc., occurring simultaneously or successively [18,19].

Due to continuous greenhouse gas emissions, climate variables and their extremes have exhibited considerable shifts across the globe [20–22]. Changes in key hydro-climatic elements and their extremes have been recorded across Canada [21,23] and unprecedented changes are projected for the future [24–26]. These changes in climate, coupled with rapid urbanization, have led to increases in the frequencies and magnitudes of flooding events in Canada. A total of 241 flood disasters have been recorded in Canada between 1990 and 2005 [4], many of which have occurred in major Canadian cities such as: Montreal in 2012, Thunder Bay in 2012, Calgary in 2013 and 2010, Winnipeg in 1997 and 2009, and Toronto in 2005 and 2013 [27]. [The trends in hydrological extremes for 248 Reference Hydrometric Basin Network (RHBN) catchments in Canada were examined by [28]. A decreasing trend in annual maximum flows for catchments located in southern Canada, and an increasing trend in catchments located in northern Canada was obtained. In addition, a robust signal of increases in spring snowmelt driven peak flow was obtained in the months of March and April, whereas a decrease in June month peak flow was obtained. These findings highlight that the behavior of extreme floods has changed across Canada as a consequence of climate change. Therefore, as advocated in previous research [29,30], it is important to obtain reliable flood frequency estimates under a non-stationary climate, such that they can be used to design climate resilient civil and municipal infrastructure across Canada.

General Climate Models (GCMs) simulate complex bio-geophysical and chemical processes occurring within the earth system and their interactions [20]. Land surface schemes are the interface within the GCMs that host important energy budget and water balance calculations occurring within a GCM grid-cell. GCM simulations are performed at a coarse spatial resolution of ~110–550 km, which hinders the accurate representation of some of the important physical processes, such as convection that shapes the earth's climate [22]. As a result, large uncertainties have been obtained in GCM projections, especially for variables linked to precipitation [22]. For making future flows and flooding projections at catchment(s) scales, typically, coarse resolution climate projections from GCMs are downscaled, and they are used to generate streamflow responses using a hydrologic model. This approach has been adopted in a number of catchment scale studies, including [25,31–34], among others.

Another approach commonly adopted by studies making future flow projections at continental or global spatial scales involve the use of coarse scale gridded runoff projections made by GCMs, and downscaling them to obtain higher resolution runoff estimates. Examples of studies adopting this approach include [35] where 45 km resolution streamflow forecasts for northeastern parts of Canada were generated by dynamically downscaling hydro-climatic forecasts from CanESM2 GCM using a CRCM4 Regional Climate Model [36]. A modified version of the WATROUTE hydraulic modelling scheme [37] was used within the CRCM4 model to simulate high resolution flows. Future changes in projected flood hazard across Europe were estimated by [38]. Dynamically downscaled future climatic projections from two regional climate models (RCMs): HIRHAM model of the Danish Meteorological Institute [39] and the Rossby Centre Atmosphere Ocean Model (RCAO) of the Swedish Meteorological and Hydrological Institute [40], were used as inputs into a hydrological model: LISFLOOD [41] to simulate 5 km resolution river discharges across Europe for historical (1961–1990) and future (2071–2100) timelines. The climatic simulations and projections from 21 GCMs were used as inputs into a global scale hydrologic model, Mac-PDM.09 [42] to simulate current and future flow regimes at 0.5° spatial resolution and assessed global water scarcity in future [43]. On the other hand, [44] used runoff simulations from GCMs and simulated high-resolution water level dynamics across the Amazon River basin, using a catchment-based macro-scale floodplain model: CaMa-Flood [45]. The same model was used by [46] to obtain 25 km resolution flow projections across the globe, using coarser resolution runoff projections from 11 GCMs in accordance with representative concentration pathways (RCPs): RCP 2.6, RCP 4.5, RCP 6.0, and RCP 8.5 [47].

This study investigates changes in the frequencies and timings of large floods (referred to address 100-year and 250-year return period flooding events hereafter) across Canada under projected future influences of climate change. The analysis presented generates novel information, as only a handful of studies (predominantly global assessments) preceding this study have assessed changes in flood hazards on a Canadian scale. In most of these studies, only changes in flooding frequencies and magnitudes have been assessed. This study extends the assessment to also analyze projected changes in flood timings, which is an important piece of information that is required for effective flood risk management. Finally, this study takes into consideration a larger ensemble of future runoff projections as compared to previous studies, which means that the results generated from this study account for the uncertainty associated with future runoff projections made by GCMs more effectively than the previous studies.

2. Models and Methods

2.1. CaMa-Flood Hydrodynamic Model

CaMa-Flood [44–46,48,49] is a global scale-distributed hydrodynamic model that routes input runoff generated by a land surface model to oceans or inland seas along a prescribed river network map. Water storage is calculated at every time-step, whereas variables such as: water level, inundated area, river discharge, and flow velocity, are diagnosed from the calculated water storage. River discharge and flow velocity are estimated using a local inertial equation. Floodplain inundation is modelled by taking into consideration the sub-grid scale variabilities in the river channel and floodplain topography. The parameters used in CaMa-Flood model are listed in Table 1. A river channel reservoir has three parameters: channel length (L), channel width (W), and bank height (B). The floodplain reservoir has a parameter for unit catchment area (A_c), and a floodplain elevation profile that describes the floodplain water depth D_f as a function of the flooded area, A_f. The topography-related parameters i.e., surface altitude (Z), distance to downstream cell (X), and unit catchment area (Ac) are calculated using the Flexible Location of Waterways (FLOW) method [50]. Finally, a Manning's roughness coefficient parameter (n) is used to represent the roughness in the river channel.

Table 1. Parameters in the catchment-based macro-scale floodplain (CaMa-Flood) model.

S. No	Name	Symbol	Unit
1	Channel length	L	m
2	Channel width	W	m
3	Bank height	B	m
4	Surface altitude	Z	m
5	Distance to downstream cell	X	m
6	Unit catchment area	A_c	m^2
7	Manning's roughness coefficient	n	$m^{-1/3}/s$

The CaMa-Flood model has been validated extensively for its ability to simulate runoff in the largest catchments of the globe [44,51]. For instance [45] evaluated the performance of the CaMa-Flood model in simulating flow characteristics in 30 major river basins, including the Amazon, Mississippi, Parana, Niger, Congo, Ob, Ganges, Lena, and Mekong. The model was found to be able to simulate flood inundation characteristics in these basins well. Furthermore, [49] evaluated the importance of adding a new computational scheme to help support the representation of flows through bifurcation channels in CaMa-Flood. The model was found to be able to perform realistic hydrodynamic calculations in the complex, resulting in mega delta with numerous bifurcation channels. Given the high credibility of the model in simulating river flow and flood inundation dynamics, the model has been used to assess the impacts of climate change at regional to global scales [46,51–55]. This study uses the globally calibrated version of the CaMa-Flood model that was used to generate global scale runoff projections in [46].

2.2. Methodology

2.2.1. Generation of 25 km Resolution Historical and Future Flows across Canada

Coarse resolution historical (1961–2005) and future (2061–2100) runoff simulations obtained from GCMs were used as inputs into the CaMa-Flood model calibrated at 25 km resolution and flow estimates covering the entire Canadian landmass. A spin-up period of two years was considered, and flows generated during this period were ignored during the assessment.

2.2.2. Grid-Wise Estimation of Future Flooding Frequencies of Historical 100- and 250-Year Floods

Generated historical and future flows at each 25 km grid were used to estimate future changes in the frequencies of historical 100- and 250-year floods. Generalized extreme value (GEV) distribution was fitted to the historical annual maximum flow series. The cumulative distribution function of the GEV distribution is expressed in Equation (1):

$$G(q) = \text{Prob } (Q \leq q) = \begin{cases} \exp[-(1-\kappa(\frac{q-\varepsilon}{\alpha}))^{\frac{1}{\kappa}}] & if\ K \neq 0 \\ \exp[-\exp(-\frac{q-\varepsilon}{\alpha})] & if\ K = 0 \end{cases} \tag{1}$$

where Q is the random variable, q is a probable value of Q, κ is the shape parameter, ε the location parameter, and α is the scale parameter. Parameters of GEV distribution were estimated using the method of L-moments [56].

Flow quantiles corresponding to 100- and 250-year return period floods were estimated for historical timelines. GEV distribution was then fitted to the annual maximums of the future flow series, and return periods corresponding to historical 100- and 250-year flood magnitudes were estimated.

2.2.3. Aggregation and Uncertainty Assessment of Projected Changes in Flooding Frequencies

Future flood frequency projections from different GCMs (corresponding to a particular emission scenario) were aggregated, and uncertainty magnitudes were quantified for each 25 km grid. Previous studies have found large uncertainties in GCM simulated projections of precipitation-related variables, with even the sign of change being found to be uncertain in many regions of the globe (IPCC 2013). Therefore, in this study, robustness of flood frequency projections was taken into consideration when aggregating flooding frequency estimates. The term 'robust' was used in this paper to refer to projections/projected changes that concurred by more than 50% of the projections analyzed. If equal numbers of projections conveyed increases/decreases in flooding frequencies in the future for a particular grid, then that grid was associated with 'non-robust' flood frequency projections. When aggregating future flood frequencies, projections that concurred with the robust sign of change in flooding frequencies were considered for aggregation and uncertainty assessment. Aggregated flooding frequencies were calculated by finding the median of future return period values, whereas uncertainty was quantified using Equation (2):

$$U_r = \frac{RP_{r,0.75} - RP_{r,0.25}}{RP_{r,0.50}} \tag{2}$$

where U_r denotes the calculated uncertainty magnitude, and $RP_{r,0.75}$, $RP_{r,0.50}$, and $RP_{r,0.25}$ denote the 75th, 50th, and 25th quantiles of the robust flood frequency projections, respectively.

2.2.4. Estimation of Historical and Future Flood Timing

When assessing changes in flood timing, flow events exceeding the 95th quantile flow value over the entire time-period of concern were considered as flooding events. For each individual CaMa-Flood grid located within Canada, flooding events were identified for both historical and future time-periods that met this criterion. The month of the year corresponding to which the largest frequency of flooding events were simulated was recorded as the time of flooding. The differences in the time of flooding between historical and future time-periods were analyzed to identify the impact of climate change on flood timing across Canada.

2.2.5. Aggregation of Historical and Future Flood Timing

Flood timing values were aggregated at each 25 km grid by only taking into consideration projections from GCMs that were able to accurately simulate the robust flow month. The robust flow month is regarded as the month that is projected with the largest number of flooding events by more than 50% of the projections analyzed. If none of the months have been concurred upon by more than 50% of the projections for a grid, then that grid is marked as having 'non-robust' flood timing results. The aggregated results for historical and future timelines are compared, to assess changes in flood timing between the two timelines.

3. Study Region

In this study, assessment of future changes in flooding frequencies and timings is performed across the entire Canadian landmass. Canada consists of 10 provinces and 3 territories: Yukon (YK), Northwest Territories (NT), Nunavut (NV), British Columbia (BC), Alberta (AB), Saskatchewan (SK), Manitoba (MB), Ontario (ON), Quebec (QB), Newfoundland and Labrador (NL), New Brunswick (NB), Nova Scotia (NS), and Prince Edward Island (PEI).

Different regions of Canada exhibit considerable differences in landscape and climate. Canada encompasses eight climate regions with different geophysical characteristics (Massey and Connors 1985). These are: (i) Pacific Maritime climate that is shaped by the presence of Pacific Ocean and is characterized by mild but extremely wet winters, and cool and dry summers. Regions located along British Columbia's west coast and its border with Yukon Territory are a part of this climate type; (ii) Cordilleran climate is influenced by continental air masses and Pacific air streams. It is found in regions covering eastern British Columbia, Yukon Territory, and small portions of southwestern Alberta. It is characterized by cold and wet winters, and warm and dry summers. The climate experienced within this climate type varies considerably spatially, because of the presence of the Rocky Mountains and insulated valleys. This climate type is found in regions covering eastern British Columbia and the Yukon Territory, as well as small portions of southwestern Alberta; (iii) Atlantic Maritime climate is influenced by western continental air masses, and it is modified by the presence of the Atlantic Ocean. This climate type is characterized by cold and wet winters and hot and wet summers. Regions encompassing New Brunswick, Nova Scotia, Prince Edward Island, and southeastern Newfoundland exhibit this climate type; (iv) Southeastern climate is influenced by the continental air masses that are modified by the presence of the Great Lakes. This climate type is characterized by cold and wet winters, and hot and wet summers. Regions that characterize this climate type include Ontario, Quebec, and parts of Nova Scotia and New Brunswick; (v) Prairies climate type is influenced by the continental air masses, and it is characterized with a wide annual temperature range with very cold winters and very hot summers; the southern regions of Alberta, Saskatchewan, and Manitoba provinces demonstrate this climate type; (vi) Boreal climate is influenced by Arctic and Pacific Ocean air masses. This climate type is characterized with very cold and dry winters, and warm and wet summers. Regions forming a continuous belt from Newfoundland and Labrador passing central Quebec and Ontario, across the Prairies, and west to the Rocky Mountains exhibit this climate type; (vii) Arctic climate region is influenced by the air stream coming from the Arctic ice pack. This region is characterized with a very harsh cold climate, permanent snow-cover, short cool summers, and minimal precipitation. Most of the Nunavut, and northern parts of Northwest Territories and Quebec exhibit this climate type. Lastly; (viii) Taiga climate region is associated with long cold winters for more than six months. This climate has some precipitation in summer and very low precipitation in winter.

4. Data Used

GCM-simulated daily runoff data for historical (1961–2005) and future (2061–2100) timelines were obtained from Coupled Model Inter-comparison Project Phase 5 (CMIP5) of the World Climate Research Programme (WCRP) (Taylor et al. 2012). Future runoff projections corresponding to four Representative Concentration Pathways (RCPs): RCP2.6, RCP4.5, RCP6.0, and RCP8.5 [47] were collected. The list of GCMs considered for analysis in this study is provided in Table 2. Runoff data for above mentioned timelines was collected for a total of 105 (84 future and 21 historical) realizations from the CMIP5 multi-model ensemble.

Table 2. GCM-RCP combinations for which at least one set of historical and future realizations were available in the CMIP5 multi-model ensemble. Note that for some cases; more than one realizations were available, and all of them are included for assessment in this study.

S. No.	GCM Names (Web Reference)	Institution	RCP 2.6	RCP 4.5	RCP 6.0	RCP 8.5
1	NorESM1-M https://portal.enes.org/models/earthsystem-models/ncc/noresm	Norwegian Climate Centre	✓	✓	✓	
2	MRI-ESM1 http://www.mri-jma.go.jp/Publish/Technical/DATA/VOL_64/index_en.html	Meteorological Research Institute				✓
3	MRI-CGCM3 http://www.glisaclimate.org/model-inventory/meteorological-research-institute-cgcm-version-3	Meteorological Research Institute	✓	✓	✓	✓
4	MPI-ESM-MR https://www.mpimet.mpg.de/en/science/models/mpi-esm/cmip5/	Max Planck Institute for Meteorology (MPI-M)	✓	✓		✓
5	MPI-ESM-LR https://www.mpimet.mpg.de/en/science/models/mpi-esm/cmip5/	Max Planck Institute for Meteorology (MPI-M)	✓	✓		✓
6	MIROC5 http://amaterasu.ees.hokudai.ac.jp/~fswiki/pub/wiki.cgi?page=CMIP5	Atmosphere and Ocean Research Institute (The University of Tokyo); National Institute for Environmental Studies; and Japan Agency for Marine-Earth Science and Technology	✓	✓	✓	✓
7	MIROC-ESM http://amaterasu.ees.hokudai.ac.jp/~fswiki/pub/wiki.cgi?page=CMIP5	Japan Agency for Marine-Earth Science and Technology; Atmosphere and Ocean Research Institute (The University of Tokyo); and National Institute for Environmental Studies	✓	✓	✓	✓
8	MIROC-ESM-CHEM http://amaterasu.ees.hokudai.ac.jp/~fswiki/pub/wiki.cgi?page=CMIP5	Japan Agency for Marine-Earth Science and Technology; Atmosphere and Ocean Research Institute (The University of Tokyo); and National Institute for Environmental Studies	✓	✓	✓	✓
9	INMCM4 http://dx.doi.org/10.1134/S0001433810040002X	Institute for Numerical Mathematics		✓		✓
10	GFDL-ESM2M http://data1.gfdl.noaa.gov/	Geophysical Fluid Dynamics Laboratory		✓	✓	✓
11	GFDL-ESM2G http://data1.gfdl.noaa.gov/	Geophysical Fluid Dynamics Laboratory	✓	✓	✓	✓

Table 2. Cont.

S. No.	GCM Names (Web Reference)	Institution	RCP 2.6	RCP 4.5	RCP 6.0	RCP 8.5
12	GFDL-CM3 http://data1.gfdl.noaa.gov/	Geophysical Fluid Dynamics Laboratory	✓	✓		✓
13	FGOALS-g2 http://www.lasg.ac.cn/fgoals/index2.asp	LASG; Institute of Atmospheric Physics; Chinese Academy of Sciences; and CESS; Tsinghua University	✓	✓		✓
14	CSIRO-Mk3-6-0 https://data.csiro.au/dap/search?q=&p=1&rpp=25&tm=Oceanography%20not%20elsewhere%20classified&sb=RELEVANCE&dr=all&collectionType=Data&topics.raw=Climate%20Change%20Processes	Commonwealth Scientific and Industrial Research Organisation in collaboration with the Queensland Climate Change Centre of Excellence	✓	✓	✓	✓
15	CNRM-CM5 https://portal.enes.org/models/earthsystem-models/cnrm-cerfacs/cnrm-cm5	Centre National de Recherches Meteorologiques / Centre Europeen de Recherche et Formation Avancees en Calcul Scientifique	✓	✓		✓
16	CMCC-CMS http://www.glisaclimate.org/node/2241	Centro Euro-Mediterraneo per I Cambiamenti Climatici		✓		✓
17	CMCC-CM https://www.cmcc.it/models/cmcc-cm	Centro Euro-Mediterraneo per I Cambiamenti Climatici		✓		✓
18	CMCC-CESM https://portal.enes.org/models/earthsystem-models/cmcc/c-esm	Centro Euro-Mediterraneo per I Cambiamenti Climatici				✓
19	CanESM2 https://www.google.com/url?q=http://climate-modelling.canada.ca/climatemodeldata/cgcm4/CanESM2/index.shtml&sa=D&ust=1516232596583000&usg=AFQjCNGO-4mT9kpaLCUnf3bpt2znikHaPw	Canadian Centre for Climate Modelling and Analysis	✓	✓		✓
20	BCC-CSM-1-1 http://forecast.bcccsm.cma.gov.cn/web/channel-34.htm	Beijing Climate Center; China Meteorological Administration	✓	✓	✓	
21	BCC-CSM-1-1-M http://forecast.bcccsm.cma.gov.cn/web/channel-34.htm	Beijing Climate Center; China Meteorological Administration	✓	✓	✓	

To perform computationally extensive CaMa-Flood simulations over the Canadian domain for all 105 realizations, the Shared Hierarchical Academic Research Computing Network (SHARCNET) platform (www.sharcnet.ca) was used. The preparation of CaMa-Flood inputs and the processing of results was performed in R programming language [57].

5. Results and Discussion

This section presents the results obtained from the assessment of projected future changes in flood frequency and timing across Canada.

5.1. Projected Changes in Flooding Frequencies

Future flooding frequencies of historical 100-year and 250-year flooding events aggregated using the approach defined in Section 2 (referred as robust GCM median approach hereafter) are presented in Figures 1 and 2 respectively. The regions presented in blue (brown) are projected with future increases (decreases) in flooding frequencies, whereas the regions presented in green are projected with no considerable changes in flooding frequencies. Regions where non-robust (described in Section 2) projections of flood frequencies are obtained are shown in white.

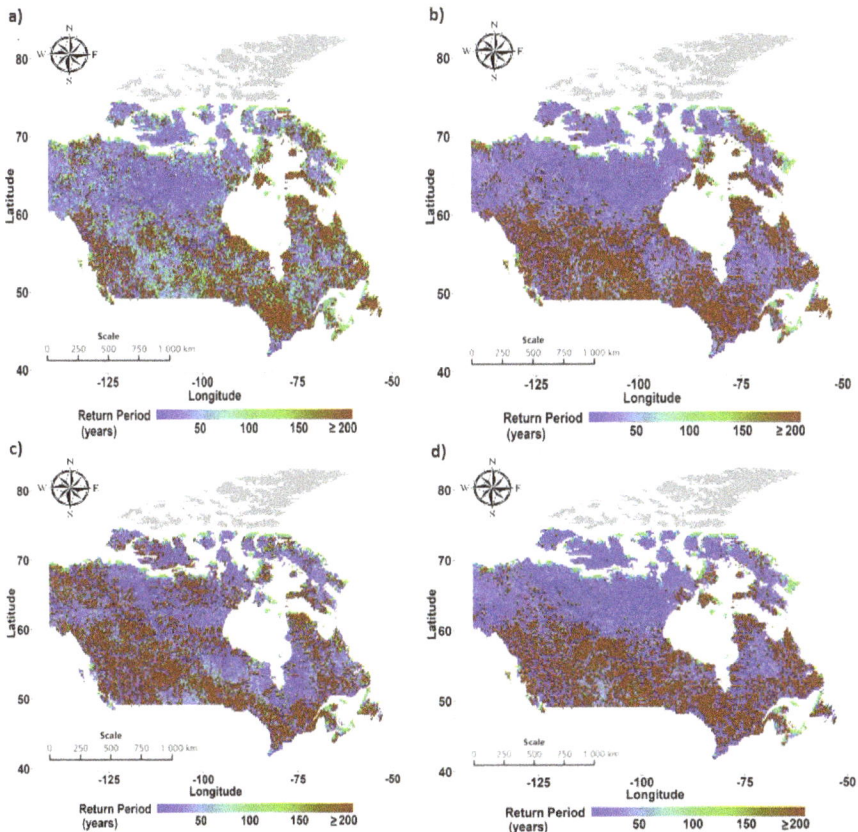

Figure 1. Future return periods of historical 100-year flood event for: (**a**) RCP 2.6; (**b**) RCP 4.5; (**c**) RCP 6.0; and (**d**) RCP 8.5. The results presented are GCM projections aggregated using the robust GCM median approach.

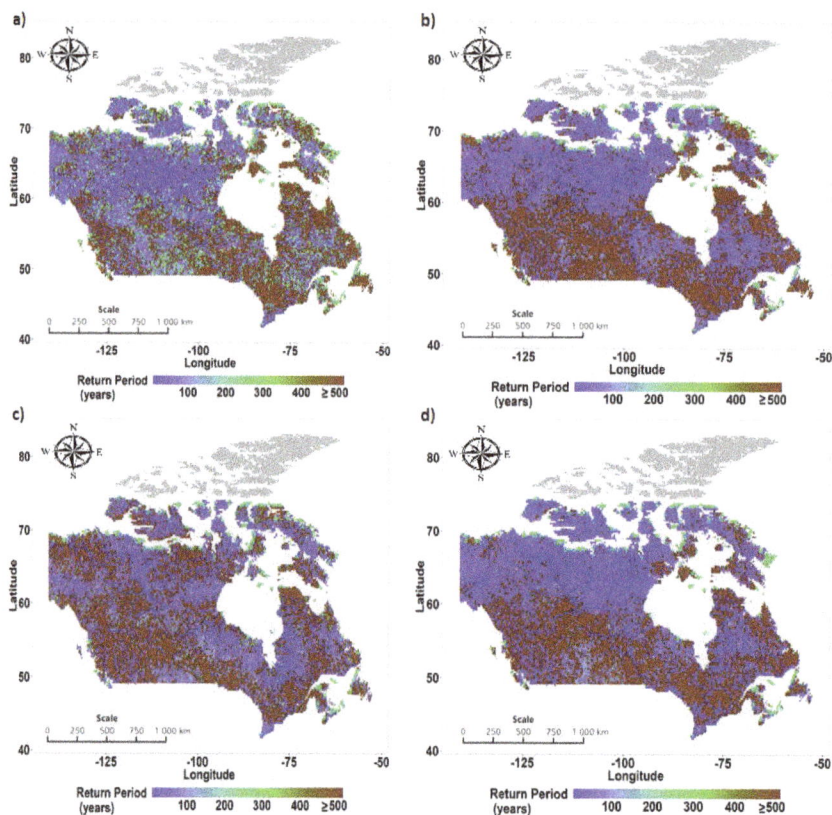

Figure 2. Future return periods of historical 250-year flood event for: (**a**) RCP 2.6; (**b**) RCP 4.5; (**c**) RCP 6.0; and (**d**) RCP 8.5. The results presented are GCM projections aggregated using the robust GCM median approach.

Results indicated that flooding frequencies of historical 100-year and 250-year return period flooding can be expected to increase considerably in the northern regions of Canada, with return periods of historical 100-year floods projected to reduce to 50-year floods or less in the future. This finding is in line with the findings from studies such as [28] which have analyzed observational flow records across Canada and have detected an increasing trend in extreme flows in the northern regions of Canada. A robust signal of projected decreases in flooding frequencies can also be noted from the results for the central and prairies regions of Canada, including areas of British Columbia, Alberta, Manitoba, and Saskatchewan, where the return period of historical 100-year floods can be expected to increase to 165–200 years in the future. These results are again in line with the observed decreases in peak flow trends in the prairies region [58] attributed largely to decreases in snowfall and increases in temperatures during the winter months. Finally, small parts of Nova Scotia, Newfoundland, and Labrador, and northernmost regions of Nunavut, and south-west British Columbia are also projected to experience no considerable changes in flooding frequencies in the future. A comparison of changes projected for 100-year and 250-year return period flooding events indicate that the spatial structure of projected changes is similar for flooding events of both magnitudes.

The spatial distribution of the projected changes in flooding frequencies was found to be similar under the RCP 4.5 and RCP 8.5 emission scenarios, whereas different spatial structures of projected changes were obtained under emission scenarios RCP 2.6 and RCP 6.0. For example, under RCP 6.0, the provinces of Yukon Territory, Northwest Territory, and Nunavut were projected with lower flood frequencies than that projected under the RCP 4.5 and RCP 8.5 emission scenarios. It should however be noted that the total number of GCMs from which runoff projections were available under RCP 6.0 (10) was lower than other emission scenarios, i.e., RCP 2.6 (15), RCP 4.5 (19), and RCP 8.5 (19), which can also contribute towards some of these differences. Finally, changes projected under RCP 2.6 were found to be of the smallest magnitudes as compared to other emission scenarios, with large areas projected with negligible changes in the future.

The above results are based on an approach where the robustness of GCMs in predicting the projected sign of change in runoff is taken into consideration when aggregating the projected changes. To investigate the impact of aggregation procedure method on the obtained results, a relatively straightforward method that does not consider the robustness of projections is used to aggregate them. In this method (referred as 'all GCM median approach' hereafter) the median of all projections is taken to obtain the future projected return periods across Canada. The results of future return periods of historical 100-year and 250-year return period flooding events obtained from this approach are shown in Figures 3 and 4, respectively. A comparison of Figures 1 and 2 with Figures 3 and 4 highlight the important similarities and differences in the magnitudes, and the spatial distributions of future projected flood frequencies. It is noted that northern parts of Canada, southwestern Ontario, and northeastern Quebec are projected with an increase in flooding frequencies when the results are aggregated from either approach. Similarly, the northern prairies region and north-central Ontario are projected with decreases in flooding frequencies from either approach. A key difference is obtained in the magnitudes of the projected changes, where higher magnitudes of absolute (positive or negative) changes were obtained from a robust GCM median approach as compared to the GCM median approach. This is likely because projected changes cancel out when averages are taken across all GCMs in the GCM median approach. For the same reason, aggregation using the all GCM median approach was found to result in more areas with no considerable changes as compared to the robust GCM median approach.

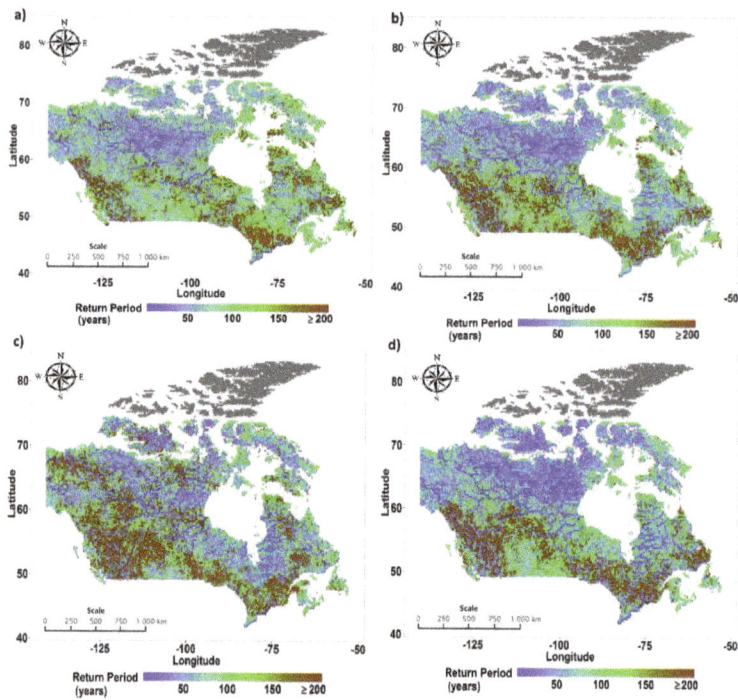

Figure 3. Future return periods of historical 100-year flood event for: (**a**) RCP 2.6; (**b**) RCP 4.5; (**c**) RCP 6.0; and (**d**) RCP 8.5. The results presented are GCM projections aggregated using the robust GCM median approach.

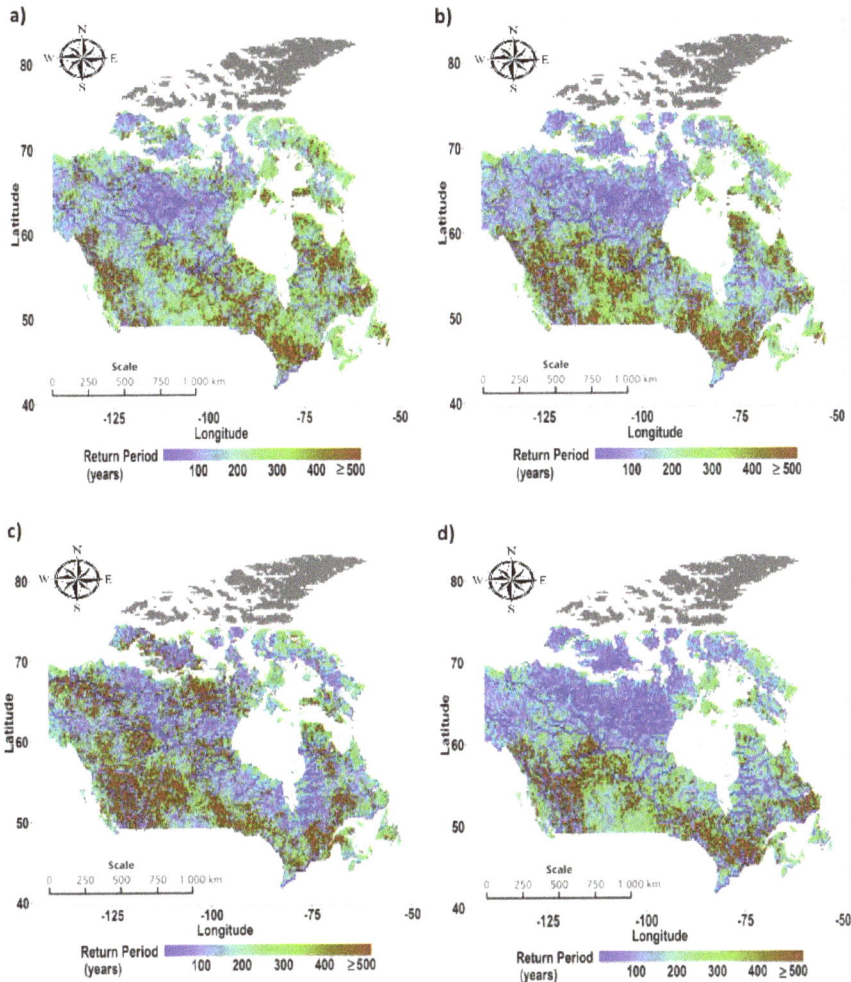

Figure 4. Future return periods of a historical 250-year flood event for: (**a**) RCP 2.6; (**b**) RCP 4.5; (**c**) RCP 6.0; and (**d**) RCP 8.5. The results presented are GCM projections aggregated using the robust GCM median approach.

Uncertainty magnitudes obtained from future flood frequency projections of historical 100-year floods obtained in the cases of the robust GCM median approach and the GCM median approach are presented in Figures 5 and 6, respectively. To present the spatial heterogeneity of uncertainty effectively, normalized values of uncertainty magnitudes are presented in the figures. The spatial distribution of uncertainty from both approaches was found to be similar; however, the uncertainty magnitudes obtained from all GCM median approaches were found to be higher than that obtained from the robust GCM median approach. Overall, increases in flood magnitudes projected in the northern provinces of Yukon, Northwest Territories, and Nunavut, northern Quebec, and south-west Ontario were found to be among the least uncertain results obtained. The decreases in flood frequency projected for the prairies region, northern Ontario, British Columbia, and Newfoundland and Labrador were found to be among the most uncertain results. These results indicated that there was a larger degree of confidence in the projected increases in flooding frequencies in parts of Canada than the projected decreases. Finally, between the four RCPs, the least uncertainty was found to be associated with the projected changes made under RCP 4.5 as compared to the other RCPs.

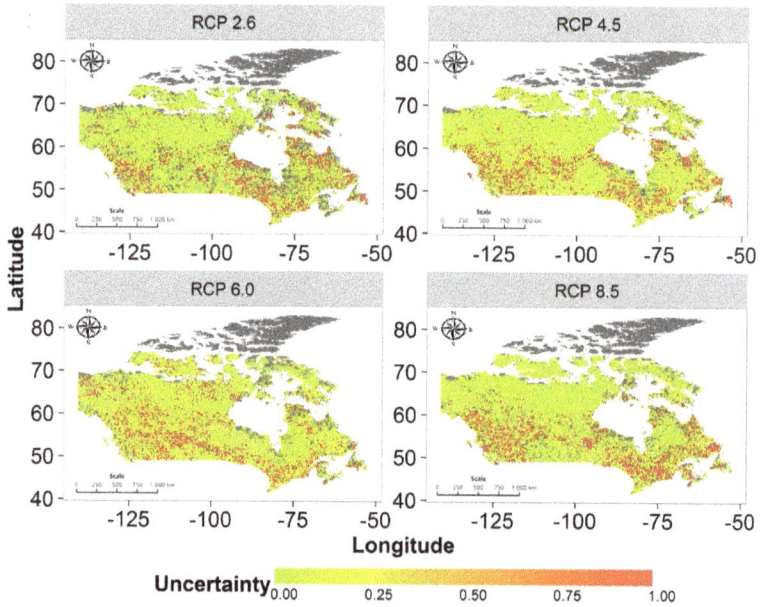

Figure 5. Normalized values of uncertainties obtained for different regions of Canada for 100-year return period flooding events when GCM projections are aggregated using a robust GCM median approach.

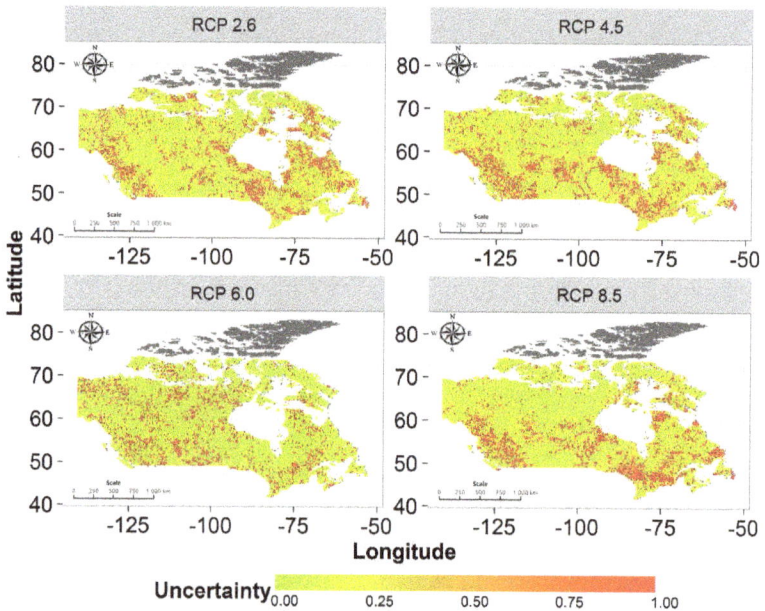

Figure 6. Normalized values of uncertainties obtained for different regions of Canada for 100-year return period flooding event when GCM projections are aggregated using a GCM median approach.

5.2. Projected Changes in Flood Occurrence Timing

Spatial distributions of flood timing obtained for historical and future timelines are presented in Figures 7 and 8 for two extreme RCPs with regard to greenhouse gas emissions: RCP 2.6 and RCP 8.5, respectively. In the figures, months where wintertime precipitation is likely to contribute to peaks, i.e., November, December, January, February, are shown in the shades of pink, months where snow-melt can be a dominant factor, i.e., March, April, May, are shown in the shades of blue, while months where summertime convection can be a dominant contributor to peaks, i.e., June, July, August, September, October, are shown in shades of green. Grids with non-robust flood timing results are shown in a grey color. Results clearly highlighted projected future increases in the total area effected by snowmelt driven floods (shown in the shades of blue), as well as an earlier onset of snowmelt driven floods in the future. These changes were most evident in the northern and central regions of Canada. Regions in Ontario and Quebec were projected with earlier summertime extreme flows (shifts from April/May to March). Most of the regions from Nunavut and Yukon Territories were projected to have earlier summertime extreme flow changes (from May to April). An earlier onset of snowmelt driven floods in the future was also evident from Figure 9, where grids that are projected with up to two months of early spring melt are shown. These results are in line with the findings from observational studies performed in different locations across Canada, where an earlier onset of snowmelt driven floods has been documented [59–62], as well as projected under the influences of climate change [34,63–65]. This observation is obtained consistently across all four emission scenarios considered for assessment, although it is noted that the GCMs were more uncertain on the prediction of the peak flow month in the cases of RCP 4.5 and RCP 8.5, than in the cases of RCP 2.6 and RCP 6.0.

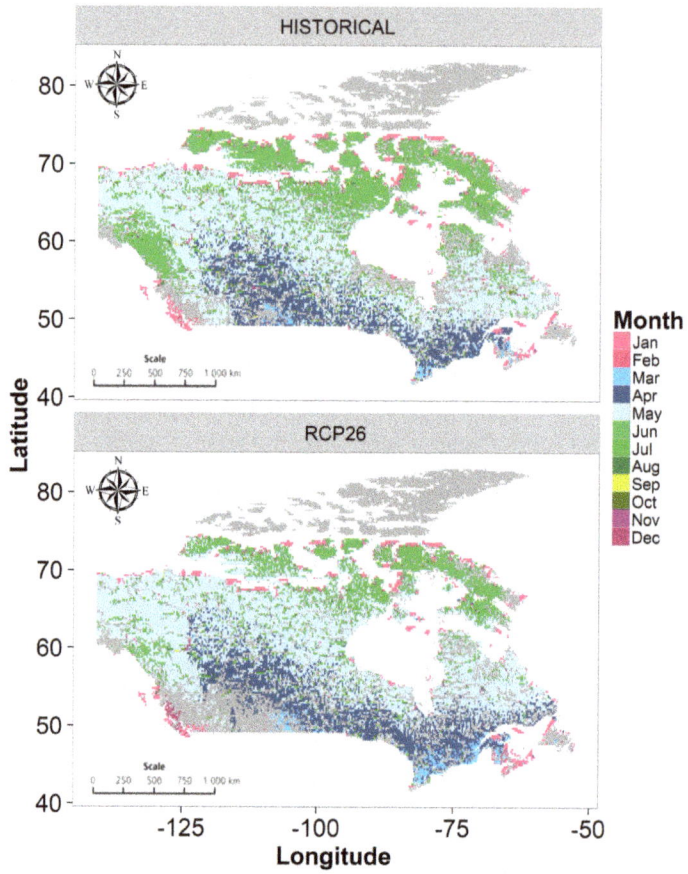

Figure 7. Monthly distribution of extreme flows obtained from aggregated runoff results obtained for historical and future timelines. Future projections are presented for RCP 2.6.

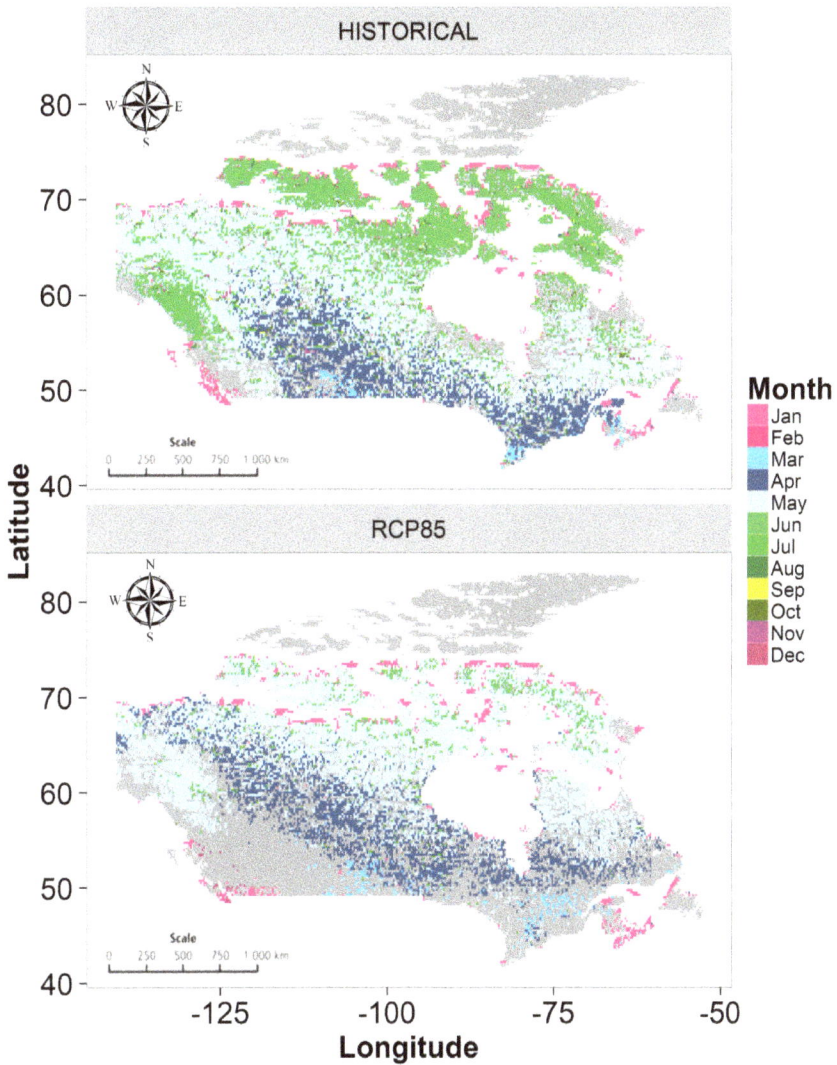

Figure 8. Monthly distribution of extreme flows obtained from aggregated runoff results obtained for historical and future timelines. Future projections are presented for RCP 8.5.

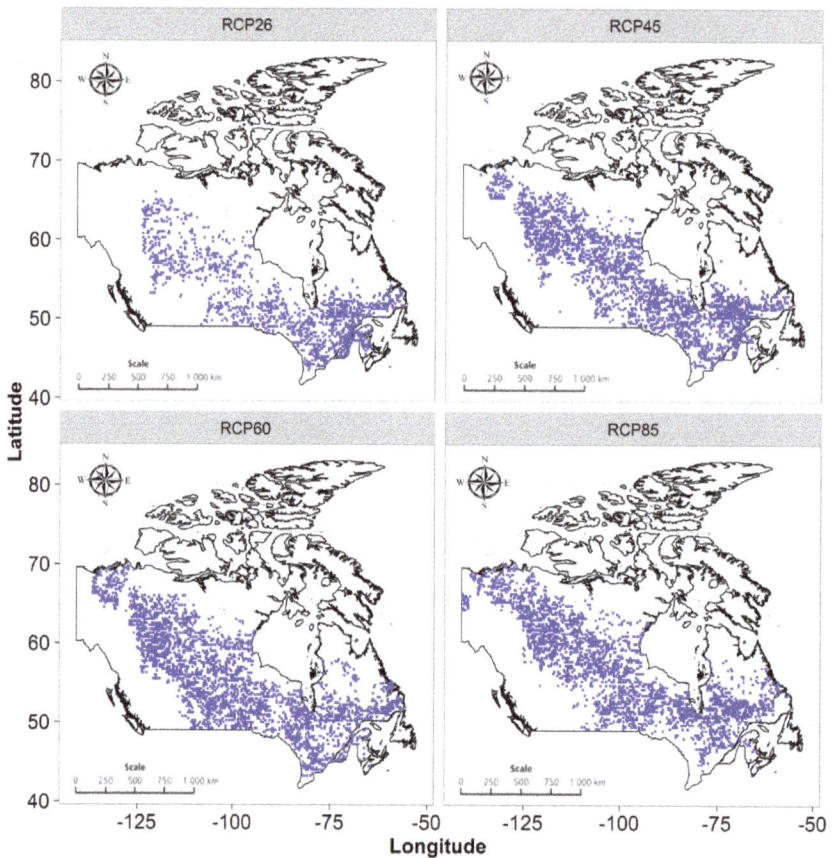

Figure 9. Grids that are projected with up to two months of earlier onset of spring-time extreme flow.

6. Conclusions

The impacts of climate change have been detected on the characteristics of streamflow and their extremes in catchments distributed across Canada. Given this non-stationarity in climatic conditions, there is a need to quantify the projected future impacts of climate change on flow extremes to better design civil and municipal infrastructure in Canada. It is also important to account for sources of uncertainties when making future projections, so that policymakers can review them before national flood protection guidelines are put into place.

This study quantifies future changes in the frequencies and timings of flooding events across Canada as a consequence of climate change. An ensemble of 84 future runoff projections made by 21 GCMs are considered for assessment. A state-of-the-art mesoscale hydrodynamic model: CaMa-Flood is used to simulate 25 km resolution historical and projected future flows from coarse resolution GCM runoff estimates. The changes projected by different GCMs are aggregated, and associated uncertainty is quantified using two approaches: (1) where only projections made by 'robust' GCMs (i.e., those who concur on the sign of change as projected by more than 50% of the GCMs) are considered, and (2) where projections made by all GCMs are considered. In general, it was found that the spatial distribution of the projected changes is similar in the results obtained from both approaches, whereas the magnitudes (both positive and negative) are found to be larger in the first approach than in the second approach. In terms of uncertainty, both approaches demonstrate similar spatial structures; however, results from the second approach demonstrate higher uncertainties than the first approach.

The spatial distributions of projected flood frequency changes convey that the northern provinces of Canada: Northwest Territories, Yukon Territory, and Nunavut, and southwestern Ontario, can be expected to have higher flood frequencies in the future, with a return period of 100-year historical floods becoming 10–60 years by the end of the 21st century. On the contrary, the northern prairies and north-central Ontario can be expected to experience lower flood frequencies, with a return period of 100-year historical floods to become 160–200 years in the future. This projected increase (decrease) in flooding frequencies in the above-mentioned areas is also found to be among the least (most) uncertain changes projected for Canada, indicating that there is a high confidence that flood hazard will increase in the above-stated regions of Canada in the future.

An assessment of projected changes in future flood timing indicates earlier snowmelt in almost all regions of Canada. This is expected, given that future temperatures are projected to increase across Canada under the influence of climate change [23]. Signs of increases in snowmelt-driven floods, and earlier snowmelt have been detected in historical flow records [66,67], as well as have been projected for many Canadian rivers [34,63–65]. The results obtained are thus also in line with the findings made in many previous studies performed at catchment scales in Canada.

The flood hazard and risk changes identified in this study can serve as useful guides for decision-makers in Canada to identify flood-hazard areas, and to prioritize appropriate mitigation and response efforts in the face of global climate change. This work can be expanded by overlapping generated flood-hazard maps with exposure elements such as population and water resource management infrastructure, to identify flood risk areas. Efforts in this direction are currently underway.

Author Contributions: A.G. conceptualized the research, performed the formal analysis, and wrote the first draft of the paper. A.G. and S.P.S. provided feedback on the research approach, and reviewed and edited the first draft of the paper. All authors revised the paper and agreed on the final version of the paper.

Funding: Funding for this research came from Chaucer Syndicates (London, UK) and the Natural Sciences and Engineering Research Council of Canada (NSERC).

Acknowledgments: We extend special thanks to Dai Yamazaki for providing a calibrated CaMa-Flood model for the purposes of this study, and for providing useful feedback on the research approach. Detailed feedback and reviews from three anonymous reviewers also helped to improve the quality of this manuscript.

Conflicts of Interest: The authors declare no conflict of interest.

References

1. Paprotny, D.; Sebastian, A.; Morales-Napoles, O.; Jonkman, S.N. Trends in flood losses in Europe over the past 150 years. *Nat. Commun.* **2018**, *9*, 1985. [CrossRef] [PubMed]
2. Paprotny, D.; Vousdoukas, M.I.; Morales-Napoles, O.; Jonkman, S.N.; Feyen, L. Compound flood potential in Europe. *Hydrol. Earth Syst. Sci. Discuss.* **2018**. [CrossRef]
3. Berghuijs, W.R.; Aalbers, E.E.; Larsen, J.R.; Trancoso, R.; Woods, R.A. Recent changes in extreme floods across multiple continents. *Environ. Res. Lett.* **2017**, *12*, 114035. [CrossRef]
4. Sandink, D.; Kovacs, P.; Oulahen, G.; McGillivray, G. *Making Flood Insurable for Canadian Homeowners*; Institute for Catastrophic Loss Reduction & Swiss Reinsurance Company Ltd.: Toronto, ON, Canada, 2010.
5. Mangini, W.; Viglione, A.; Hall, J.; Hundecha, Y.; Ceola, S.; Montanari, A.; Rogger, M.; Salinas, J.L.; Borzi, I.; Parajka, J. Detection of trends in magnitude and frequency of flood peaks across Europe. *Hydrol. Sci. J.* **2018**, *63*. [CrossRef]
6. Dandapat, K.; Panda, G.K. A geographic information system-based approach of flood hazards modelling; Paschim Medinipur district; West Bengal; India. *J. Disaster Risk Stud.* **2018**, *10*, 518. [CrossRef] [PubMed]
7. Zischg, A.P.; Felder, G.; Weingartner, R.; Quinn, N.; Coxon, G.; Neal, J.; Freer, J.; Bates, P. Effects of variability in probable maximum precipitation patterns on flood losses. *Hydrol. Earth Syst. Sci.* **2018**, *22*, 2759–2773. [CrossRef]

8. Paprotny, D.; Morales-Napoles, O.; Jonkman, S.N. Efficient pan-European river flood hazard modelling through a combination of statistical and physical models. *Nat. Hazards Earth Syst. Sci.* **2017**, *17*, 1267–1283. [CrossRef]

9. Parkes, B.; Demeritt, D. Defining the hundred year flood: A Bayesian approach for using historic data to reduce uncertainty in flood frequency estimates. *J. Hydrol.* **2016**, *540*, 1189–1208. [CrossRef]

10. Li, C.; Cheng, X.; Li, N.; Du, X.; Yu, Q.; Kan, G. A Framework for Flood Risk Analysis and Benefit Assessment of Flood Control Measures in Urban Areas. *Int. J. Environ. Res. Public Health* **2016**, *13*, 787. [CrossRef] [PubMed]

11. Iacobellis, V.; Castorani, A.; Santo, A.R.D.; Gioia, A. Rationale for flood prediction in karst endorheic areas. *J. Arid Environ.* **2015**, *112A*, 98–108. [CrossRef]

12. Herget, J.; Roggenkamp, T.; Krell, M. Estimation of peak discharges of historical floods. *Hydrol. Earth Syst. Sci.* **2014**, *18*, 4029–4037. [CrossRef]

13. McSharry, P.E.; Little, M.A.; Rodda, H.J.; Rodda, J. Quantifying flood risk of extreme events using density forecasts based on a new digital archive and weather ensemble predictions. *Q. J. R. Meteorol. Soc.* **2013**, *139*, 328–333. [CrossRef]

14. Keast, D.; Ellison, J. Magnitude Frequency Analysis of Small Floods Using the Annual and Partial Series. *Water* **2013**, *5*, 1816–1829. [CrossRef]

15. Fiorentino, M.; Gioia, A.; Iacobellis, V.; Manfreda, S. Regional analysis of runoff thresholds behaviour in Southern Italy based on theoretically derived distributions. *Adv. Geosci.* **2011**, *26*, 139–144. [CrossRef]

16. Reis, D.S.; Stedinger, J.R. Bayesian MCMC flood frequency analysis with historical information. *J. Hydrol.* **2005**, *313*, 97–116. [CrossRef]

17. Blazkov, S.; Beven, K. Flood frequency prediction for data limited catchments in the Czech Republic using a stochastic rainfall model and TOPMODEL. *J. Hydrol.* **1997**, *195*, 256–278. [CrossRef]

18. Moftakhari, H.R.; Salvadori, G.; AghaKouchak, A.; Sanders, B.F.; Mathews, R.A. Compounding effects of sea level rise and fluvial flooding. *Proc. Natl. Acad. Sci. USA* **2016**, *114*, 9785–9790. [CrossRef] [PubMed]

19. Lin, N.; Kopp, R.E.; Horton, B.P.; Donnelly, J.P. Hurricane Sandy's flood frequency increasing from year 1800 to 2100. *Proc. Natl. Acad. Sci. USA* **2016**. [CrossRef] [PubMed]

20. IPCC. Summary for Policymakers. In *Climate Change. The Physical Science Basis. Contribution of Working Group I to the Fifth Assessment Report of the Intergovernmental Panel on Climate Change*; Stocker, T.F., Qin, D., Plattner, G.-K., Tignor, M., Allen, S.K., Boschung, J., Nauels, A., Xia, Y., Bex, V., Midgley, P.M., Eds.; Cambridge University Press: Cambridge, UK; New York, NY, USA, 2013.

21. IPCC. *Managing the Risks of Extreme Events and Disasters to Advance Climate Change Adaptation. A Special Report of Working Groups I and II of the Intergovernmental Panel on Climate Change*; Field, C.B., Barros, V., Stocker, T.F., Qin, D., Dokken, D.J., Ebi, K.L., Mastrandrea, M.D., Mach, K.J., Plattner, G.K., Allen, S.K., Eds.; Cambridge University Press: Cambridge, UK; New York, NY, USA, 2012.

22. Prein, A.F.; Rasmussen, R.M.; Ikeda, K.; Liu, C.; Clark, M.P.; Holland, G.J. The future intensification of hourly precipitation extremes. *Nat. Clim. Chang.* **2017**, *7*, 48–52. [CrossRef]

23. ECCC (Environment and Climate Change Canada). Climate Data and Scenarios for Canada: Synthesis of Recent Observation and Modelling Results. 2016. Available online: https://ec.gc.ca/sc-cs/default.asp?lang=En&n=80E99404-1&printfullpage=true&wbdisable=true#wb-info (accessed on 12 September 2018).

24. Gaur, A.; Eichenbaum, M.K.; Simonovic, S.P. Analysis and modelling of surface Urban Heat Island in 20 Canadian cities under climate and land-cover change. *J. Environ. Manag.* **2017**, *206*, 145–157. [CrossRef] [PubMed]

25. Mandal, S.; Simonovic, S.P. Quantification of uncertainty in the assessment of future streamflow under changing climate conditions. *Hydrol. Processes* **2017**, *31*, 2076–2094. [CrossRef]

26. Mladjic, B.; Sushama, L.; Khaliq, M.N.; Laprise, R.; Caya, D.; Roy, R. Canadian RCM Projected Changes to Extreme Precipitation Characteristics over Canada. *J. Clim.* **2011**, *24*, 2565–2584. [CrossRef]

27. Sandink, D. Urban Flooding in Canada. *Inst. Catastr. Loss Reduct.* **2013**, *52*, 1–94.

28. Burn, D.H.; Hag Elnur, M.A. Detection of hydrological trends and variability. *J. Hydrol.* **2002**, *255*, 107–122. [CrossRef]

29. Salas, J.D.; Obeysekera, J. Revisiting the Concepts of Return Period and Risk for Nonstationary Hydrologic Extreme Events. *J. Hydrol. Eng.* **2014**, *19*, 554–568. [CrossRef]

30. Milly, P.C.D.; Betancourt, J.; Falkenmark, M.; Hirsch, R.M.; Kundzewicz, Z.W.; Lettenmaier, D.P.; Stouffer, R.J. Stationarity Is Dead: Whither Water Management? *Science* **2008**, *319*, 573–574. [CrossRef] [PubMed]
31. Gaur, A.; Simonovic, S.P. *Climate Change Impact on Flood Hazard in the Grand River Basin*; Water Resources Research Report no. 084; Facility for Intelligent Decision Support, Department of Civil and Environmental Engineering: London, ON, Canada, 2013.
32. Linde, A.H.; Aerts, J.C.J.H.; Bakker, A.M.R.; Kwadijk, J.C.J. Simulating low probability peak discharges for the Rhine basin using resampled climate modeling data. *Water Resour. Res.* **2010**, *46*, W04512. [CrossRef]
33. El-Khoury, A.; Seidou, O.; Lapen, D.R.; Que, Z.; Mohammadian, M.; Sunohara, M.; Bahram, D. Combined impacts of future climate and land use changes on discharge; nitrogen and phosphorus loads for a Canadian river basin. *J. Environ. Manag.* **2015**, *151*, 76–86. [CrossRef] [PubMed]
34. Eum, H.I.; Dibike, Y.; Prowse, T. Comparative evaluation of the effects of climate and land-cover changes on hydrologic responses of the Muskeg River; Alberta; Canada. *J. Hydrol. Reg. Stud.* **2016**, *8*, 198–221. [CrossRef]
35. Huziy, O.; Sushama, L.; Khaliq, M.N.; Laprise, R.; Lehner, B.; Roy, R. Analysis of streamflow characteristics over Northeastern Canada in a changing climate. *Clim. Dyn.* **2013**, *40*, 1879–1901. [CrossRef]
36. De-Elia, R.; Cote, H. Climate and climate change sensitivity to model configuration in the Canadian RCM over North America. *Meteorol. Z.* **2010**, *19*, 325–339. [CrossRef]
37. Soulis, E.D.; Snelgrove, K.R.; Kouwen, N.; Seglenieks, F.; Verseghy, D.L. Towards closing the vertical water balance in Canadian atmospheric models: Coupling of the land surface scheme CLASS with the distributed hydrological model WATFLOOD. *Atmos. Ocean* **2000**, *38*, 251–269. [CrossRef]
38. Dankers, R.; Feyen, L. Climate change impact on flood hazard in Europe: An assessment based on high resolution climate simulations. *J. Geophys. Res.* **2008**, *113*, D19105. [CrossRef]
39. Christensen, J.H.; Christensen, O.B.; Lopez, P.; van Meijgaard, E.; Botzet, M. *The HIRHAM4 Regional Atmospheric Climate Model*; Scientific Report 96-4; Danish Meteorological Institute: Copenhagen, Denmark, 1996.
40. Jones, C.G.; Willen, U.; Ullerstig, A.; Hansson, U. The Rossby Centre Regional Atmospheric Climate Model part I: Model climatology and performance for the present climate over Europe. *R. Swed. Acad. Sci.* **2004**, *33*, 199–210. [CrossRef]
41. De Roo, A.P.J.; Wesseling, C.G.; Van Deurzen, W.P.A. Physically-based river basin modelling within a GIS: The LISFLOOD model. *Hydrol. Processes* **2000**, *14*, 1981–1992. [CrossRef]
42. Gosling, S.N.; Arnell, N.W. Simulating current global river runoff with a global hydrological model: Model revisions; validation; and sensitivity analysis. *Hydrol. Process.* **2011**, *25*, 1129–1145. [CrossRef]
43. Arnell, N.W.; Gosling, S.N. The impacts of climate change on hydrological regimes at the global scale. *J. Hydrol.* **2013**, *486*, 351–364. [CrossRef]
44. Yamazaki, D.; Lee, H.; Alsdorf, E.; Dutra, E.; Kim, H.; Kanae, S.; Oki, T. Analysis of the water level dynamics simulated by a global river model: A case study in the Amazon River. *Water Resour. Res.* **2012**, *48*, W09508. [CrossRef]
45. Yamazaki, D.; Kanae, S.; Kim, H.; Oki, T. A physically based description of floodplain inundation dynamics in a global river routing model. *Water Resour. Res.* **2011**, *47*, 1–21. [CrossRef]
46. Hirabayashi, Y.; Mahendran, R.; Koirala, S.; Konoshima, L.; Yamazaki, D.; Watanabe, S.; Kanae, S. Global flood risk under climate change. *Nat. Clim. Chang.* **2013**, *3*, 816–821. [CrossRef]
47. Van Vuuren, D.P. The representative concentration pathways: An overview. *Clim. Chang.* **2011**, *109*, 5–31. [CrossRef]
48. Yamazaki, D.; de Almeida, G.AM.; Bates, P.D. Improving computational efficiency in global river models by implementing the local inertial flow equation and a vector-based river network map. *Water Resour. Res.* **2013**, *49*, 7221–7235. [CrossRef]
49. Yamazaki, D.; Sato, T.; Kanae, S.; Hirabayashi, Y.; Bates, P.D. Regional flood dynamics in a bifurcating mega delta simulated in a global river model. *Geophys. Res. Lett.* **2014**, *41*, 3127–3135. [CrossRef]
50. Yamazaki, D.; Oki, T.; Kanae, S. Deriving a global river network map and its sub-grid topographic characteristics from a fine-resolution flow direction map. *Hydrol. Earth Syst. Sci.* **2009**, *13*, 2241–2251. [CrossRef]
51. Ikeuchi, H.; Hirabayashi, Y.; Yamazaki, D.; Kiguchi, M.; Koirala, S.; Nagano, T.; Kotera, A.; Kanae, S. Modeling complex flow dynamics of fluvial floods exacerbated by sea level rise in the Ganges-Brahmaputra-Meghna delta. *Environ. Res. Lett.* **2015**, *10*, 124011. [CrossRef]

52. Hu, X.; Hall, J.W.; Shi, P.; Lim, W.H. The spatial exposure of the Chinese infrastructure system to flooding and drought hazards. *Nat. Hazards* **2016**, *80*, 1083–1118. [CrossRef]

53. Mateo, C.M.; Hanasaki, N.; Komori, D.; Tanaka, K.; Kiguchi, M.; Champathong, M.; Sukhapunnaphan, T.; Yamazaki, D.; Oki, T. Assessing the impacts of reservoir operation to floodplain inundation by combining hydrological, reservoir management, and hydrodynamic models. *Water Resour. Res.* **2014**, *50*, 7245–7266. [CrossRef]

54. Koirala, S.; Hirabayashi, Y.; Mahendran, R.; Kanae, S. Global assessment of agreement among streamflow projections using CMIP5 model outputs. *Environ. Res. Lett.* **2014**, *9*, 064017. [CrossRef]

55. Pappenberger, F.; Dutra, E.; Wetterhall, F.; Cloke, H.L. Deriving global flood hazard maps of fluvial floods through a physical model cascade. *Hydrol. Earth Syst. Sci.* **2012**, *16*, 4143–4156. [CrossRef]

56. Vogel, R.M.; Wilson, I. Probability distribution of annual maximum; mean; and minimum streamflows in the United States. *J. Hydrol. Eng.* **1996**, *1*, 69–76. [CrossRef]

57. R Development Core Team. *R: A Language and Environment for Statistical Computing*; R Foundation for Statistical Computing: Vienna, Austria, 2018; ISBN 3-900051-07-0.

58. Burn, D.H.; Fan, L.; Bell, G. Identification and quantification of streamflow trends on the Canadian Prairies. *Hydrol. Sci. J.* **2008**, *53*, 538–549. [CrossRef]

59. Rokaya, P.; Budhathoki, S.; Lindenschmidt, K.-E. Trends in the Timing and Magnitude of Ice-Jam Floods in Canada. *Sci. Rep.* **2018**, *8*, 5834. [CrossRef] [PubMed]

60. Semmens, K.A.; Romage, J.; Bartsch, A.; Liston, G.E. Early snowmelt events: Detection; distribution; and significance in a major sub-arctic watershed. *Environ. Res. Lett.* **2013**, *8*, 014020. [CrossRef]

61. Déry, S.J.; Stahl, K.; Moore, R.D.; Whitfield, P.H.; Menounos, B.; Burford, J.E. Detection of runoff timing changes in pluvial, nival and glacial rivers of western Canada. *Water Resour. Res.* **2009**, *45*, W04426. [CrossRef]

62. Stewart, I.T.; Cayan, D.R.; Dettinger, M.D. Changes toward Earlier Streamflow Timing across Western North America. *J. Clim.* **2005**, *18*, 1136–1155. [CrossRef]

63. Dibike, Y.; Shakibaeinia, A.; Eum, H.; Prowse, T.; Droppo, I. Effects of projected climate on the hydrodynamic and sediment transport regime of the lower Athabasca River in Alberta, Canada. *River Res. Appl.* **2018**. [CrossRef]

64. Poitras, V.; Sushama, L.; Seglenieks, F.; Khaliq, M.N.; Soulis, E. Projected Changes to Streamflow Characteristics over Western Canada as Simulated by the Canadian RCM. *J. Hydrometeorol.* **2011**, *12*, 1395–1413. [CrossRef]

65. Pohl, S.; Marsh, P.; Bonsal, B.R. Modeling the Impact of Climate Change on Runoff and Annual Water Balance of an Arctic Headwater Basin. *Arctic* **2006**, *60*, 173–186. [CrossRef]

66. Whitfield, P.H.; Cannon, A.J. Recent Variations in Climate and Hydrology in Canada. *Can. Water Resour. J.* **2000**, *25*, 19–65. [CrossRef]

67. Zhang, X.; Harvey, K.D.; Hogg, W.D.; Yuzyk, T.R. Trends in Canadian Streamflow. *Water Resour. Res.* **2001**, *37*, 987–998. [CrossRef]

water

MDPI

Article

The Influence of Flow Projection Errors on Flood Hazard Estimates in Future Climate Conditions

Joanna Doroszkiewicz [1], Renata J. Romanowicz [1,*] and Adam Kiczko [2]

[1] Institute of Geophysics Polish Academy of Sciences, Ks. Janusza 64, 01-542 Warsaw, Poland;
     joador@igf.edu.pl
[2] Faculty of Civil and Environmental Engineering, University of Life Sciences (WULS-SGGW),
     Nowoursynowska 166, 02-787 Warsaw, Poland; adam_kiczko@sggw.pl
*   Correspondence: romanowicz@igf.edu.pl; Tel.: +48-22-6915-852

Received: 16 November 2018; Accepted: 18 December 2018; Published: 29 December 2018

Abstract: The continuous simulation approach to assessing the impact of climate change on future flood hazards consists of a chain of consecutive actions, starting from the choice of the global climate model (GCM) driven by an assumed CO_2 emission scenario, through the downscaling of climatic forcing to a catchment scale, an estimation of flow using a hydrological model, and subsequent derivation of flood hazard maps with the help of a flow routing model. The procedure has been applied to the Biala Tarnowska catchment, Southern Poland. Future climate projections of rainfall and temperature are used as inputs to the precipitation-runoff model simulating flow in part of the catchment upstream of a modeled river reach. An application of a lumped-parameter emulator instead of a distributed flow routing model, MIKE11, substantially lowers the required computation times. A comparison of maximum inundation maps derived using both the flow routing model, MIKE11, and its lump-parameter emulator shows very small differences, which supports the feasibility of the approach. The relationship derived between maximum annual inundation areas and the upstream flow of the study can be used to assess the floodplain extent response to future climate changes. The analysis shows the large influence of the one-grid-storm error in climate projections on the return period of annual maximum inundation areas and their uncertainty bounds.

Keywords: flood inundation maps; climate change; EURO-CORDEX projections; continuous simulations

---

1. Introduction

Following the EU Floods Directive (2007) [1], flood risk maps should take into account climate changes and should be updated in cycle, every six years. A standard approach to deriving flood risk maps, applied in Poland, consists of running the 1D MIKE11 model for design flood waves of different return periods (e.g., 500, 100, 10 years). Water management and adaptation to floods in Poland for a 100-year flood have been planned. Design flood wave and model simulations have been applied to estimate 100-year inundation outlines [2]. In flood risk assessment, a design flood is customarily applied when flow data are available. When annual flow series are short, event-based simulation approaches can be applied [3]. This type of approach uses a rainfall generator as an input to a rainfall–runoff model. Annual maximum flow events are subsequently selected from the obtained flow series and used as inputs to a 1-D hydraulic model to derive maximum inundation maps. Due to the relatively large computation time requirements of a distributed flow routing model, a continuous simulation approach is rarely applied to flood risk assessments [4].

In order to provide flood risk maps for the future, there is a need to include future climate projections in the design of a flood wave [5,6]. The design flood is usually estimated from the probability distribution of annual maximum flows, which may be influenced by water engineering structures or urbanization [7]. Human interference at the river side changes the water balance of the river and the character of flood events. The above-mentioned factors make flood frequency analysis (FFA) based on annual peak flows under the assumption of stationarity not very robust when future climatic and human-induced changes are expected. In addition, the standard approach assumes that, for example, a 1-in-100 year flow quantile will be transformed into a 1-in-100 year inundation extent. That assumption imposes the existence of a linear transformation between the flow and an inundation area downstream, which is not usually met in practice.

The other approach, followed here, consists of a continuous simulation of a climate projection–runoff–flow routing system for an ensemble of climate projections for the entire length of available records and direct derivation of the 1-in-100 year flood inundation maps. The possible scenarios of the flood protection schemes and changes in land use (e.g., urbanization) may be taken into account [8]. This approach avoids both drawbacks of the standard method, but may require a high amount of computing time, which will increase when the uncertainty of simulations is to be included in the derived risk maps.

The important point related to flood wave design, which includes three flood wave parameters (wave height, height, length and base level), is the uncertainty. This uncertainty might undermine the usefulness of the flood risk assessment [9–12]. In addition, an analysis of the influence of different adaptation scenarios requires an application of a distributed flow routing model that takes most of the computing time in the simulation approach.

In order to facilitate the computations, we propose an emulator of the 1D MIKE11 model. An emulator is a simple model approximating particular actions of the original model and is used instead of that model for computationally intensive applications. Two main types of emulators can be distinguished: structure-based emulators, where the mathematical structure of the emulator is derived from the simplifications of the original model, and data-based emulators, where the emulator structure is identified and its parameters estimated from a dataset generated from planned experiments performed on the complex simulation model [13].

The distributed 1D model of a riverside, treated as a complex model, can be successfully solved by data-based dynamic emulation modeling within the following steps [12]: (i) parameter estimation by calibration and validation of the emulator, (ii) analysis of the uncertainty of the output, and (iii) assimilation of the data to be analyzed within the emulator with defined parameters. The results of the emulator can be treated as an input to the flood risk assessment, which constitutes the basis for scenario analysis and a contribution to the decision-making process.

Decision-making with respect to adaptation or mitigation is usually based on flood risk assessments, which can be burdened by a quite high uncertainty [14]. This uncertainty can be divided into two groups: reducible and non-reducible. The reducible uncertainty is dependent on the amount of available information applied to the model, whereas the non-reducible uncertainty comes from the randomness of nature.

In Poland, even though the EU Floods Directive (2007) commits all EU countries to follow "the flood risk adaptation cycle," the new Polish Water Act gives a high degree of freedom to meet the EU requirements and interpret the Act in different ways [15]. Since 1 January 2018, the responsibility for decision-making has been in the hands of local government. Local government bases its decision on existing master plans, and the regional flood prevention assessments usually do not contain an uncertainty analysis. At the county scale, the risk assessment seems to be no longer necessary, and the same refers to scenarios for adaptation.

The propagation of uncertainty [16,17] in the routing model begins at the stage of climate projection. Both global and regional climate projections display systematic errors [18,19]. Projections of temperature and precipitation data used as an input to a hydrological model may be over- or underestimated due to the climate model characteristics and bias related to downscaling [20,21].

The aim of this study is an assessment of the sensitivity of projections of annual maximum inundation area to climate projections and modeling errors, including rainfall–runoff and flow routing models. The analysis framed by this article in the context of Floods Directive 2007 satisfies the flood hazard assessment, which is an inherent part of flood risk assessment, enriched by the projection of future climate and uncertainty analysis but without cost–benefit analysis.

The Biala Tarnowska catchment in Southern Poland was used as a case study. Future projections of precipitation and temperature obtained from the EURO-CORDEX project were applied as inputs to the HBV model simulating flow in the upper part of the catchment, at the Ciezkowice gauging station. The sensitivity of the derived future inundation areas at Tuchow meander on the Biala River to climate projection errors was assessed.

The following content is organized within sections that present the methodology, describe the case study area, describe the emulator, discuss the results and the uncertainty of the results of both MIKE11 and the emulator of MIKE11, and apply the developed procedure to climate projections. Final comments on the achievements of this work are given in the conclusions.

2. Materials and Methods

2.1. Approach

The approach can be summarized in the following steps:

1. calibration and validation of a distributed model (MIKE11) for the chosen river reach;
2. development of an emulator for the selected cross sections of the model in the form of the input–output transformation;
3. verification of the emulator using independent observation sets;
4. comparison of the deterministic inundation extent maps obtained using both approaches;
5. estimation of prediction error bands (uncertainty analysis of a flow routing model);
6. derivation of dependence of annual maximum inundation area on the upstream flow;
7. dependence of distribution of future annual maximum inundation areas on climate model instabilities.

The first step required the development of a distributed model for a selected river reach, using the available DTM data and measured cross sections. The choice of MIKE11 was dictated by the need to compare the results of modeling to the existing, MIKE11-based flood risk maps for the region of Biala Tarnowska.

In the second step, the emulator structure was developed based on the modeling goals and the characteristics of the processes involved [22]. The emulator calibration (Step 3) applied MIKE11 simulations as a source of information on unobserved water levels at cross sections along the river reach. Verification of the emulator (Step 4) was performed using the reference data from the time period not applied in the calibration stage. A comparison of the deterministic inundation extent maps provided an insight into the quality of the emulator spatial predictions. In the next step (5), the uncertainty of emulator predictions was assessed. The emulator was treated in this step as an additional source of errors, apart from the error related to the distributed flow routing model. We investigated the differences between MIKE11 predictions and the emulator predictions for the observations at Koszyce Wielkie. However, we cannot assess errors at unobserved cross sections, unless we have additional spatial data, even in the form of historical inundation maps. It would be possible to calculate the parametric errors by sampling MIKE11 roughness parameter space and derivation in the emulator for each sample. As there are already many studies on distributed model uncertainties [23], we focused on uncertainties related to the emulator itself. We considered observations of water levels at the Koszyce Wielkie gauging station to evaluate the emulator performance and its predictive uncertainty. The sixth step involved the derivation of a functional dependence (sensitivity curve) between maximum inundation area and the upstream flow based on the calibration data. The seventh step involved a comparison of the frequency distribution of the maximum annual inundation areas based on future climate projections with and without one-grid-storms caused by regional model instabilities.

2.2. Case Study—The Biala Tarnowska Catchment

The Biala Tarnowska catchment, chosen as a case study, is a mountainous catchment in Southern Poland. The catchment is semi-natural, and one of the semi-natural Polish catchments, chosen following an extensive analysis of available hydro-meteorological and geomorphological data [24]. The catchment, with an area of 966.9 km², has a forest-covered upper part and a lower part covered mainly by agricultural lands (Figure 1).

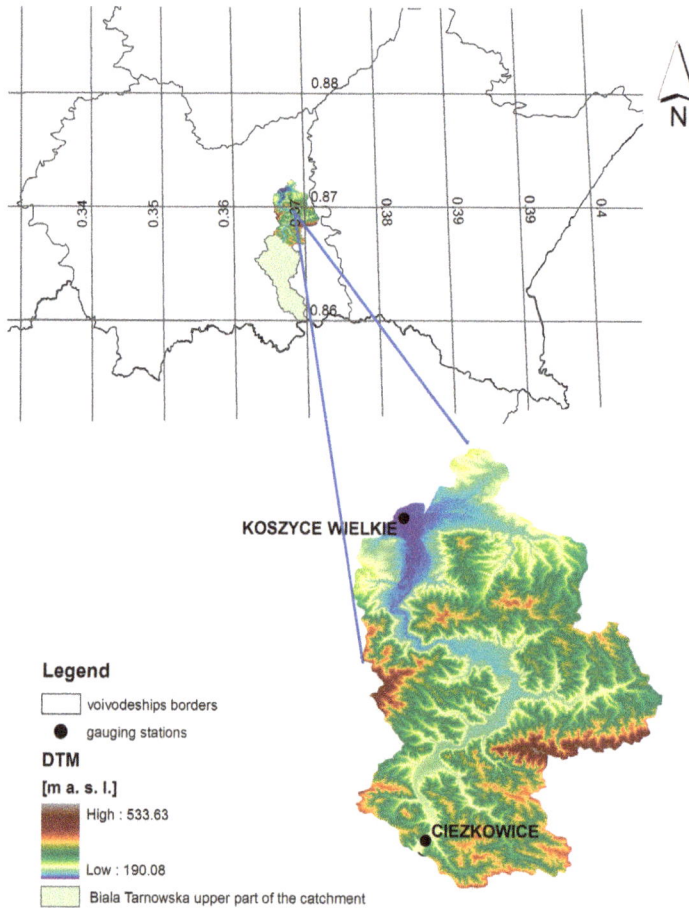

Figure 1. The location of the study catchment; left picture: county scale, whole catchment; right picture: local scale, routing model part.

The upper part of the catchment, down to the Ciezkowice gauging station (527.5 km²) with 97% classified as forest cover, was modeled by the HBV rainfall–runoff model.

Thiessen polygons were applied to interpolate point precipitation data for the purpose of rainfall–runoff model calibration and verification. The lower part of the river between Ciezkowice and Koszyce Wielkie gauging stations was modeled using the 1-D MIKE11 flow routing model and the emulator, based on a stochastic transfer function.

Daily hydro-metrological observations of precipitation, temperature, and streamflow and estimated potential evapotranspiration were used as an input to the hydrological model HBV [25]. Observed historical hydrological and climate daily time series of precipitation, temperature, and streamflow for 40 years from November 1970 to October 2010 were obtained from the National Water Resources and Meteorological Office (IMGW) in Poland. Daily potential evapotranspiration was calculated using the temperature-based Hamon approach [26]. Daily streamflow data from the Ciezkowice and Koszyce Wielkie hydrological stations for a period of 40 years (1971–2010) were used in the calibration (1971–2000) and validation (2001–2010) stages.

In the case of Biala Tarnowska, observations at the upstream gauging station Ciezkowice and the downstream, Koszyce Wielkie, show that one recording observations daily (6 a.m.) might miss the fast changes of flow in short flood events. The delay between observations in upstream and downstream is 8 h. Because of this, an additional check-up of input daily data was done by a comparison of 10 min observation data and interpolated daily data. The correlation between them was 0.93, which may lead to the conclusion that 1 h output results will contain the peaks of flood events.

2.3. Routing Model for Biala Tarnowska River

Two parallel approaches to modeling future inundation areas are presented schematically in Figure 2. The scheme shows the simulation chain of actions necessary to produce flood adaptation indices. At the left-hand side of Figure 2, the chain of actions starts from the input data (observed or simulated daily temperature and precipitation series), which are pre-processed for the whole time horizon of the simulations. They form the input to a hydrological model, which provides flow simulations subsequently routed using the MIKE11 model. The adaptation actions may consist of building the polder release at parts of the river reach to prevent downstream-located cities from flooding, building embankments, introducing levees, or other engineering constructions.

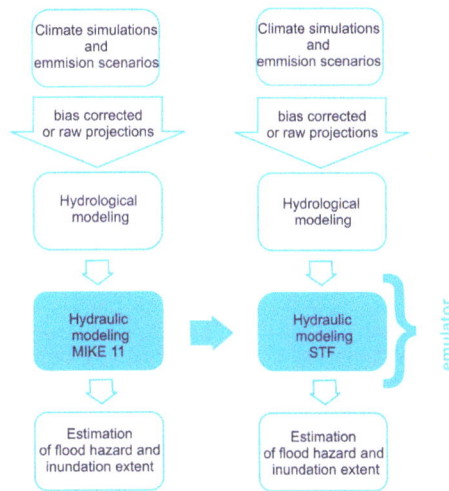

Figure 2. Scheme of an application of an emulator of MIKE11 for climate projections.

Distributed models, e.g., MIKE11 [27], SOBEK [28], and HEC-RAS [29], use the logic of a real-world system and follow the laws governing the routing of the river. Distributed flow routing is needed to test different adaptation measures and their influence on flood risk. However, modeling and comparison of the outcomes of flood adaptation measures require multiple simulations of the entire system, presented in the left panel of Figure 2.

The application of MIKE11 in this context requires a high amount of computing time. The approach proposed here consists of replacing the MIKE11 model by a transfer function-based emulator as shown in the right panel of Figure 2. The development of the emulator is based on the distributed model simulations, and its derivation is described in the following sub-sections.

The national climate change adaptation strategy for Poland, in the case of adaptation to floods, recommends the application of a top-down flow routing software—MIKE11 (DHI, Hørsholm, Denmark). For the sake of comparison of results, the same software has been chosen in this study.

The routing model of the Ciezkowice–Koszyce Wielkie reach of the Biala Tarnowska River includes 118 cross sections specified using the triple zone approach.

The triple zone approach defines the water depth border value and the resistance in each cross section. All cross sections were built using direct field measurements, and values were generated from the digital elevation model (DTM). The resolution of DTM is 1 m horizontally and 0.01 m vertically.

Due to the fact that, in the Biala Tarnowska reach, the water balance depends on ungauged tributaries in almost 50% of the total flow, the model was supplied with 19 point sources representing the tributaries and one lateral source. In our approach, we followed the discussion on lateral sources in the routing model presented in [30]. The authors show the impact of the saturated zone on matrix–conduit exchanges in a shallow phreatic aquifer and highlight the important role of the unsaturated zone on storage. In this work, flows of ungauged tributaries were estimated based on the HBV results for the whole Biala Tarnowska catchment and the participation of the tributary catchment areas in the Ciezkowice–Koszyce Wielkie part of the catchment. The flows do not take into account the delays caused by saturation but are simply weighted by the size of ungauged tributary sub-catchments.

The flow routing model has been simulated with a 1 min time step interval, using the HBV simulated flow time series for the period 1996–2000 as an upstream boundary condition and the rating curve as the downstream boundary. The *fmincon* optimisation routine from MATLAB® was applied to optimize six MIKE11 roughness parameters. The Nash–Sutcliff index (NS) was used to describe the goodness of model fit for the observed flows and water levels at the Koszyce Wielkie gauging station. The values of the NS index in Koszyce Wielkie, using flows as a criterion for the calibration and validation periods, were 0.80 and 0.77, respectively, whilst the NS indices for water levels were 0.60 and 0.54, respectively.

2.4. Input—The Hydrological Model for the Ciezkowice Gauging Station

In this study, the HBV hydrological model has been applied to the upper catchment of Biala Tarnowska (down to the Ciezkowice gauging station). The model applies daily temperature, evaporation and precipitation data as an input. The HBV [25] model parameters were optimized using the DEGL minimization code [31]. The parameters optimized by the DEGL are given in Table 1.

Table 1. The HBV model parameter values for the Biala catchment, Ciezkowice and Koszyce Wielkie.

HBV Routines	Name of the Parameter	Ciezkowice	Koszyce Wielkie	Unit
soil routine	P.FC	61.19	119.55	mm
	P.BETA	1.68	2.89	non
	P.LP	1	1	non
	P.CFLUX	0.6	0.23	mm/day
surface water balance routine	P.ALFA	0.24	0.27	non
	P.KF	0.3	0.11	day
	P.KS	0.11	1.64	day
soil moisture routine	P.PERC	1.33	1.12	mm/day
snow routine	O.TT	1.23	1.05	°C
	O.TTI	4.93	7	°C
	O.FOCFMAX/DTTM	0.01	1.36	°C/mm day
	O.CFMAX	1.27	1.00	°C/mm day
	O.CFR	0.71	0.00	non
	O.WHC	0	0	mm/mm

In addition, the HBV model was also used for the entire catchment closing at the Koszyce Wielkie gauging station to estimate the ungauged tributaries to the Biala reach between Ciezkowice and Koszyce Wielkie. The optimal model parameters for the whole catchment were obtained from [32] and are presented in Table 1.

2.5. The Structure of the MIKE11 Emulator

The emulator consists of a Box–Cox transform and a stochastic transfer function (STF) model [33]. The choice of the Box–Cox-STF-based model was dictated by its transparent structure combining nonlinear input transformation of the input variable (Box–Cox) with linear model dynamics (STF), forming the so-called Hammerstine-type model [34].

The emulator parameters are derived using the distributed model simulations, available at each model cross section and discrete in time. For the purpose of the derivation of flood inundation maps for future climate projections, the model input should have the form of flows, and model output should be in the form of water levels. Assuming that the emulator's single module is built between the 1st and *n*th cross section, with the flow at an upstream end of the reach acting as an input variable, and the water levels at a cross section acting as an output variable, the system's basic module has the structure as shown in Figure 3.

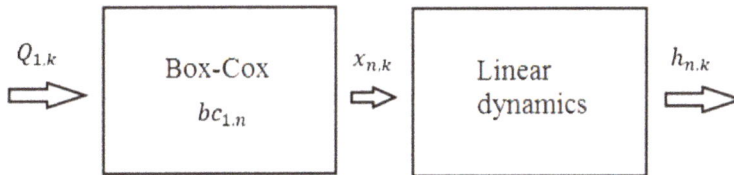

Figure 3. Module of an emulator based on flow–water-level relationship; $Q_{1,k}$ denotes flow at the 1st cross section at time k, for $k = 1, \ldots, T$; $h_{n,k}$ denotes water levels at the *n*th cross section at time k; $bc_{1,n}$ denotes the Box–Cox transformation between flow used as an input at the first cross section and the *n*th transfer function input $x_{n,k}$.

The emulator equation for the *n*th cross section has the form of a single-input single-output (SISO) model with the Box–Cox transformation of an input:

$$y_{n,k} = \frac{R_n}{(1 - P_n \cdot z^{-1})} x_{k - \delta_n} + \varepsilon_k \tag{1}$$

$$x_k = \frac{(Q_{1,k} + c_n)^{b_n} - 1}{b_n} \tag{2}$$

where $y_{n,k}$ denotes water levels at an *n*th *MIKE11* cross section, $x_{n,k}$ denotes transformed by the Box–Cox model flow $Q_{1,k}$ at the Ciezkowice cross section, and ε_k is the $N(0, \sigma)$ error. Parameters R_n, P_n, c_n, b_n, and δ_n are specific for each cross section.

The optimization of parameters of a Box–Cox transform was performed jointly with the STF model parameters, using the NS criterion to take into account the model output bias. From this point of view, the procedure was similar to the derivation of a nonlinear gain on the input in the flow forecasting model [34]. Its output can be expressed either in the form of flow or water levels at required spatial locations (e.g., the Koszyce Wielkie reach of the river).

It should be noted that any other suitable simplified flow routing model can be used as an emulator. The emulator with the STF describing process dynamics has the advantage of providing well-defined model parameters with a known covariance structure as part of its output. The nonlinear Box–Cox static transformation of flow is used to describe a nonlinear flow–water-level relationship and to improve the estimation of STF model parameters.

The emulator of a distributed flow routing model applied here has the form of parallel connected STF-type modules. The number of modules will depend on the number of cross sections required to build the inundation map, as shown in Figure 4. In the case study, 118 cross sections of the Biala River were used as components of the emulator. The part of the Ciezkowice–Koszyce reach includes the Tuchow meander, 16 km long, which will be used later as an illustration. It is shown by black squares in Figure 4 and consists of 38 cross sections. The emulator was optimized with the fminsearch function (part of the MATLAB function library), which applies the Nelder–Mead simplex method [35]. Both the NS index and R_T^2 were applied as objective functions evaluated at the vertices of a simplex. The adaptive nature of the process requires continuous revision of simplexes to find the best fit to the response surface. Each module is calibrated using the outputs (here the water levels at each cross section) obtained from the fully distributed model. The form of a module depends on the variable chosen to describe the system dynamics [22].

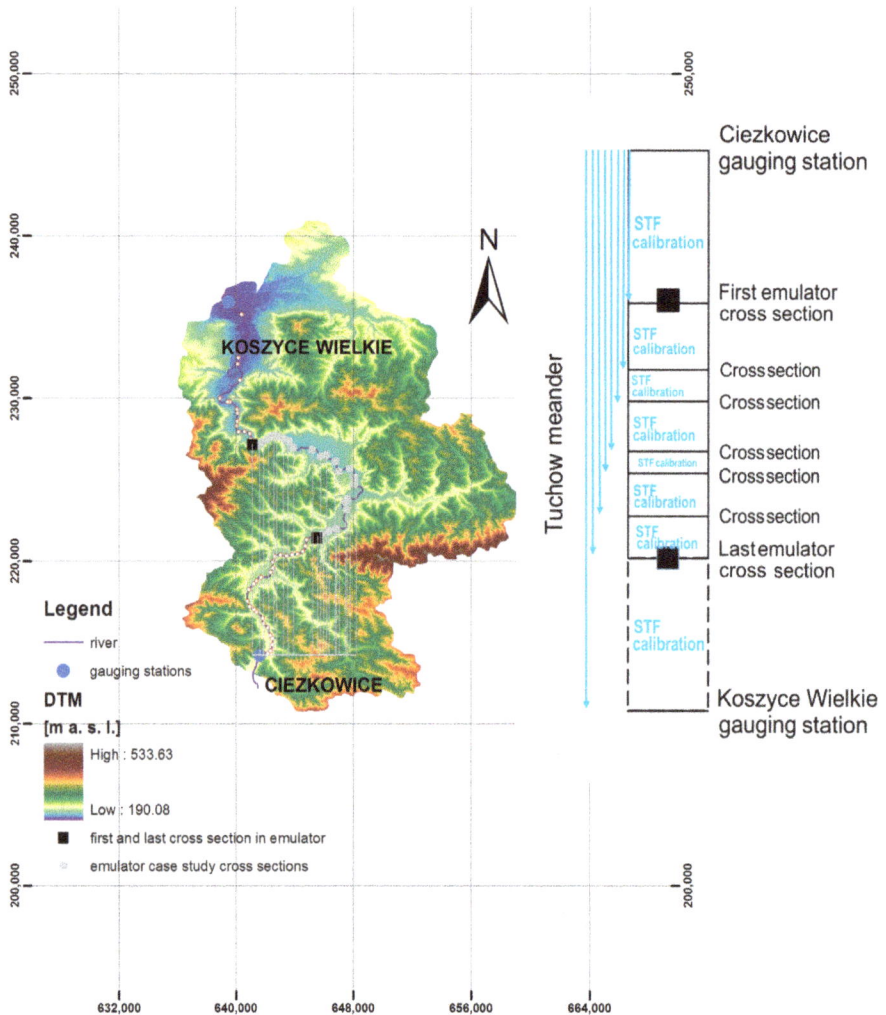

Figure 4. The STF-based MIKE11 emulator of water level predictions; on the Biala Tarnowska River—the meander near Tuchow city.

To take into account flood wave durations less than 24 h, hourly time steps were used for the estimation of emulator parameters. In the calibration stage, the best model fit in R_T^2 varied between 0.97 and 0.98, which shows a very good emulator performance. Besides the goodness of fit, the other parameters characterize the emulator. The pure advection delay between the first and the last Tuchow meander cross section is approximately 2 h, depending on the wave speed. However, the total time of wave propagation in the emulator from the beginning of the model to the last cross section of the meander was estimated at between 5 and 9 h [36]. The emulator well reproduces MIKE11 simulations, which is not surprising, as the input–output relationship between the distributed model variables is well defined. The accuracy of the validation expressed by the coefficient of determination R_T^2 is of a similar order as in the calibration stage (about 0.97). The total accuracy of the emulator predictions depends also on the MIKE11 accuracy, which can be tested at only one available gauging station: Koszyce Wielkie.

The results of the validation of water level predictions at the Koszyce Wielkie gauging station are presented in Figure 5. The emulator predictions are shown by a continuous black line together with 0.95 confidence bands depicted by the shaded area. The MIKE11 predictions are shown by the black dotted line, and the observed water levels are shown by the red line.

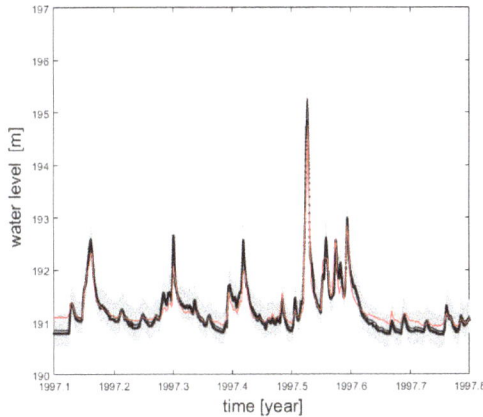

Figure 5. The STF-based MIKE11 emulator of water level predictions, validation stage, the Ciezkowice–Koszyce Wielkie model. The black dots denote MIKE11 simulations; the continuous black line denotes the emulator predictions; the red line denotes the observed water levels at the Koszyce Wielkie gauging station; the shaded area denotes 0.96 confidence limits.

The SISO model could be introduced for Biala Tarnowska despite the tributaries present in the catchment due to the fact that they are strongly correlated with the main river flow. In the case where tributaries are not strongly correlated, the MISO (multiple-input single-output) approach is required [34].

3. Results

3.1. Comparison of Deterministic Flood Inundation Maps

Flood inundation areas derived from the water level simulations of MIKE11 and the emulator are compared in the following subsection. The relationships between the annual maximum flow at Ciezkowice, forming the input to the emulator (Q) and corresponding water levels (H) and inundation area (A) at the Tuchow meander simulated by the emulator, are presented in Figure 6. The left-hand panel shows the H(Q) relationship for the 10th meander cross section, the middle panel shows A(Q), and the right-hand panel shows the relationship between the water level at the 10th meander cross section and the meander annual maximum inundation area A(H). The correlation coefficients of H(Q), A(Q), and A(H) are 0.96, 0.98, and 0.99, respectively. In the A(H) plot (Figure 6, right hand panel), it can be seen that, for an increasing water depth value with a constant value of 1 m (dashed lines), the gradient of the inundated area changes. At a 5 m water depth, the embankments are exceeded, causing changes in the inundation area response.

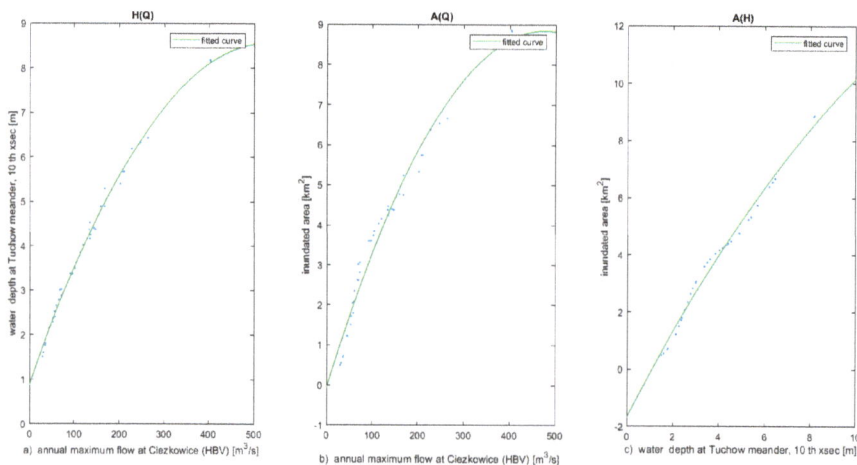

Figure 6. Dependence on hydrological input (annual maxima) and routing model predictions for the Tuchow meander for the reference period 1971–2010. (**a**) Water depth values at the 10th cross section of the meander versus the maximum annual flow; (**b**) inundation areas at the Tuchow meander versus the maximum annual flow; (**c**) inundation area versus the maximum annual water depth at the 10th cross section of the Tuchow meander. Blue dots denote the data, and green lines denote the fitted curve to the square polynomial.

The above illustration shows that relationships between the maximum inundation areas and inflow are nonlinear (middle panel of Figure 6). They can be used to derive local sensitivity indices describing how the inundation area responds to the variation in input flow at Ciezkowice for adaptation purposes. In particular, the A(Q) relationship depends on the channel and floodplain geometry. In this case, the inundation area is more sensitive for smaller maximum flow values than for larger values.

The computation time of the emulator is much shorter than that of MIKE11. As illustrated in Table 2, the time of preparation of the inundation map using the emulator shows a difference of two orders of magnitude.

Table 2. Time processing for the 1971–2010 period and an area of 42 km^2.

Time of Processing	Map Based on MIKE11 Results (s)	Map Based on the Emulator of MIKE11 Results (s)
Routing	19,035	147
calculating a flooded area and saving the inundation map based on 1 m × 1 m resolution DTM	990	990
TOTAL	20,015 (≈5 h 34 min)	1137 (≈19 min)

The vast amount of time might be consumed by mapping the water levels on the DTM. The time of mapping depends on the size of the area covered and its resolution, and it substantially influences the total computation time.

Figure 7 presents differences in the extent of maximum inundation area for the year 1997 for the lower part of the Tuchow meander. The emulator shows about 3.5% larger inundation area than MIKE11. The critical success index (CSI) was used [37,38] to assess the differences quantitatively. The CSI assumes that the times when an event is not expected and not observed have no consequence. The equation is as follows:

$$CSI = \frac{observed\ event}{(observed\ event+simulated\ event\ but\ not\ observed+observed\ event\ but\ not\ simulated)} \qquad (3)$$

In this investigation, an observed event is the inundation area simulated with the distributed MIKE11 model results, and the simulated event is that simulated with the results of the emulator of MIKE11 model. However, as spatial observations are missing, we do not know if MIKE11 under-predicts or over-predicts the real inundation extent.

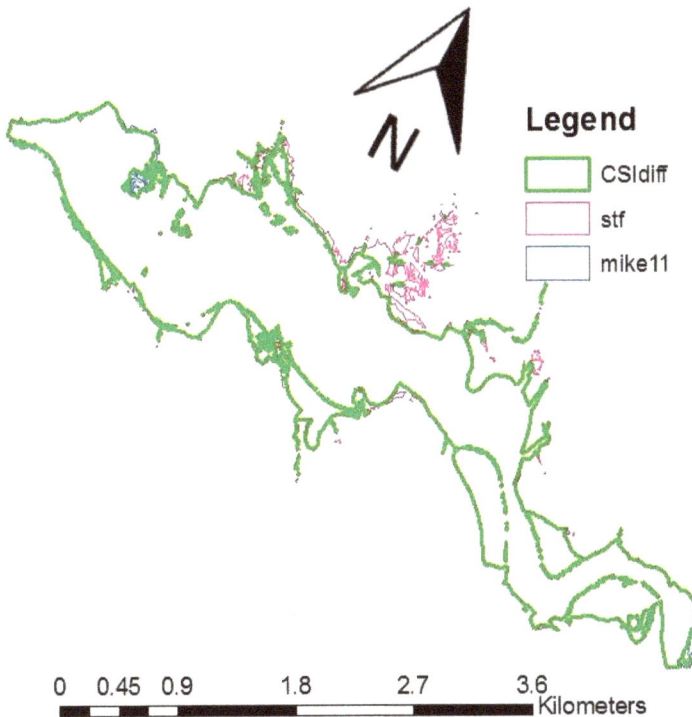

Figure 7. Maximum inundation extent of MIKE11 (blue) and the emulator of MIKE11 (magenta) for the year 1997 for a part of the Tuchow meander, and the CSI-based difference (green).

The CSI for the 1997 flood event equals 0.91. The application of the CSI eliminates grid cells that were differently classified by either of the approaches.

3.2. The Uncertainty of Emulator Predictions

The uncertainty of the emulator predictions depends on the uncertainty of the MIKE11 model predictions and on the emulator parametric uncertainties.

The uncertainty of the emulator of MIKE11 predictions can be assessed directly by comparison with the available observations. Unfortunately, no observations of either water levels or inundation extent are available along the modeled river reach, and only water levels at the gauging station in Koszyce Wielkie can be used for model validation and an estimation of the uncertainty of its predictions.

Following the standard STF model procedures, the model parameters, together with an estimate of their uncertainty and the prediction variance, were derived using the observations. When the MIKE11 model outputs were used instead of the measured values, the emulator prediction error was extended by the unknown error term, which also included the Box–Cox transform error.

Whilst the STF model parameter uncertainty is known for each MIKE11 cross section, the prediction error can be derived only for those cross sections where observations are available. Therefore, the STF model uncertainty describes the lower constraint (minimum value) of the uncertainty of the emulator predictions.

In order to estimate the influence of the Box–Cox transform and the STF model uncertainty on water levels generated by the emulator, we apply Monte Carlo simulations of the emulator parameters to generate an ensemble of possible model predictions. As each model provides the estimates of parameter uncertainty in the form of a covariance matrix, the simulations follow the information enclosed in those estimates. The procedure was applied to the Koszyce Wielkie gauging station. The model parameters and their uncertainty estimates are given in Table 3.

From the point of view of flood risk assessment, only the water level values that are above the bank-full are of interest. As the Koszyce Wielkie river reach has high embankments, the water level values that cause inundation upstream or downstream of Koszyce Wielkie are still in-bank at that cross section. In order to estimate their uncertainty related to the emulator parameters, the peak values above the 75th percentile (above 192 m) were selected from the generated ensemble of hourly water levels. The approach of [39] was followed in choosing the high-flow event series. The minimum distance between the events was set to 40 h, following the analysis of flow residence times to avoid the correlation between the events.

An example of event selection is presented in Figure 8. The correlation between the events was assessed using Bartlett's test of independence for a significance level of 5%. As a result of that test, the independence assumption was not rejected.

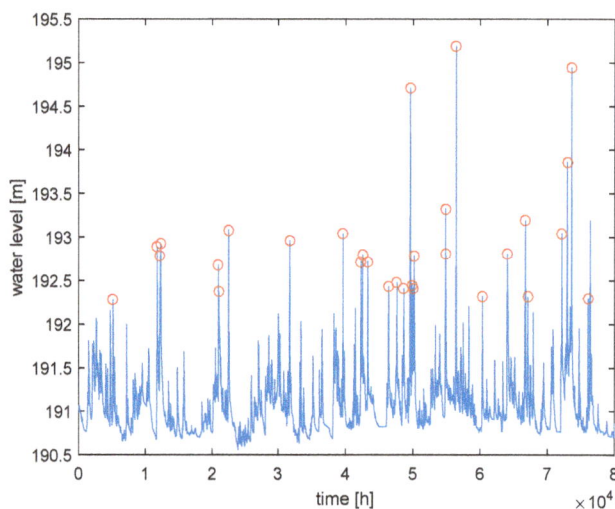

Figure 8. Classification of high water levels at Koszyce Wielkie gauging station for the threshold value equal to the 75th percentile.

The high flow events were used to derive the empirical water level-return-period curves following the approach described by [40].

The empirical return period $T(i)$ of the i-th sorted highest peak water level was calculated from the ranked peak water level series according to the formula:

$$T(i) = n/i \tag{4}$$

where n denotes the number of years in a maximum water level series.

Table 3. Parameter values of the emulator of the MIKE11 Ciezkowice–Koszyce Wielkie model and their uncertainty. Parameters are defined by Equations (1) and (2), where P/R denotes the cross-correlation between P and R.

Parameter [Unit]	Mean	Variance
P [-]	0.8710	0.1943×10^{-6}
R [-]	0.0199	0.0045×10^{-6}
P/R [-]	-	0.029×10^{-6}
δ [h]	13	0
c [-]	-0.2740	0.03
b [-]	0.5246	0.03
ε [-]	0	0.008

Figure 9 presents an example of an empirical relationship between the return period and water levels obtained for the 10-year long time series of simulated water levels at the Koszyce Wielkie gauging station for the period 1991–2000. The black-marked curves correspond to nine different parameter sets of the Monte–Carlo (MC) ensemble simulations, while the red-marked curve is obtained for the observed time series.

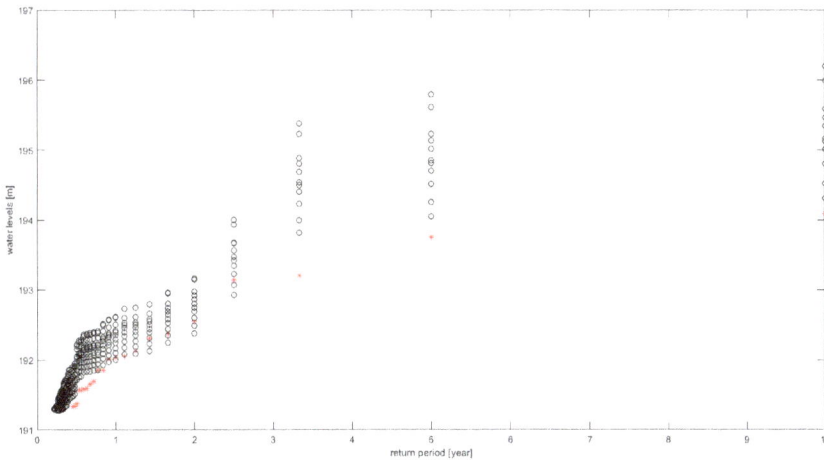

Figure 9. Example of return-period–water-level relationship for the peak water level ensemble.

The evaluation of return periods allows for a comparison of the modeling results with the observations for different return periods. Figure 10 presents the derived 1-in-5 and 1-in-10 year return period water levels in the Koszyce gauging station obtained from an MC analysis of emulator predictions. The observations, denoted by red perpendicular lines, indicate slight over-predictions of emulator simulations. The 0.95 confidence bands for the 1-in-10 year return period predictions are about 2 m wide. We noted that the Box–Cox model parameters have the largest influence on the spread of the results and can be used to derive weights describing model goodness of fit. However, those weights are local, i.e., they depend on the local differences between the predicted and modeled water levels and cannot be extrapolated to other sites.

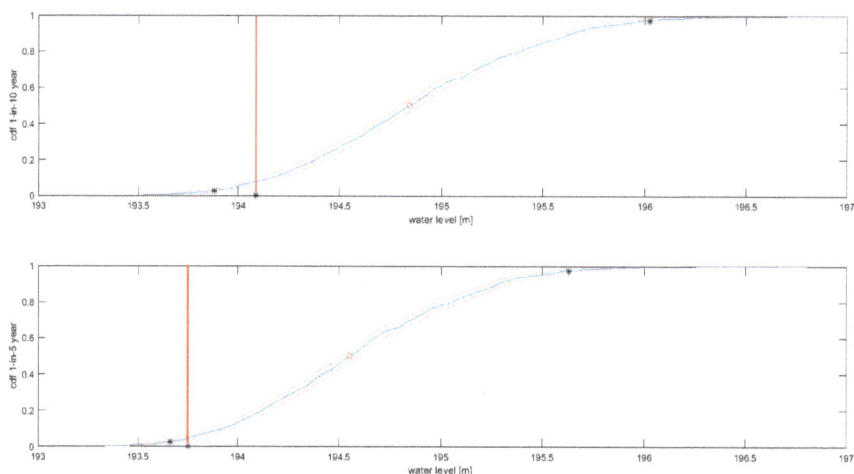

Figure 10. Empirical cumulative distribution function (cdf) of water levels with 1-in-10 year return period (upper panel) and 1-in-5 year return period (lower panel) obtained for the Monte Carlo (MC) simulations of peak water levels at Koszyce Wielkie (blue lines); red dotted lines show the cdf 0.95 confidence bounds; black stars show the 0.025, 0.5, and 0.975 quantiles water levels with 1-in-10 and 1-in-5 year return periods; the observed water levels corresponding to 1-in-10 and 1-in-5 year return periods are shown by the red lines.

This illustration indicates the importance of observations of inundation extent (water levels) that would allow the uncertainty of the predictions to be reduced.

The resulting water level uncertainty is transposed into the inundation extent uncertainty through the water-level–inundation-extent relationship (right panel of Figure 6). This will result in a different inundation area uncertainty depending on the incoming upstream flow.

3.3. Derivation of Flood Hazard Maps for the Future

The EURO-CORDEX initiative project [41] was used as a source of daily temperature and precipitation projections for the period 1971–2100. The projections have a spatial resolution of 12.5 km (EUR-11) for the RCP4.5 emission scenario. The ensemble applied in this study has seven members, consisting of four different RCMs driven by three different GCMs (Table 4).

Table 4. List of RCM/GCMs used in this study.

Model Number	GCM	RCM	Full Name	Source Institution
1	EC-EARTH	RCA	Rossby Center regional	Swedish Meteorological and Hydrological Institute
2	EC-EARTH	HIRHAM	Not applicable	Danish Meteorological Institute
3	EC-EARTH	CCLM	Community Land Model	Not applicable
4	EC-EARTH	RACMO	Regional Atmospheric Climate Model	Not applicable
5	MPI	CCLM	Community Land Model	Not applicable
6	MPI	REMO	Regional-scale Model	Max Planck Institute for Meteorology
7	CNRM	CCLM	Community Land Model	CERFACS, France

The bias correction of the projected climate variables is usually performed in order to improve their physical interpretability [42]. However, all bias correction methods alter the extreme values, which might be not justified for high flow extremes. Therefore, it was decided to use raw precipitation projections in the present study of projections of maximum flood inundation extent. The simulation procedure described in Section 2 was followed to produce a series of projected annual maximum inundation areas for the Tuchow meander for the seven climate models, for the 1971–2100 period. The results are presented in Figure 11.

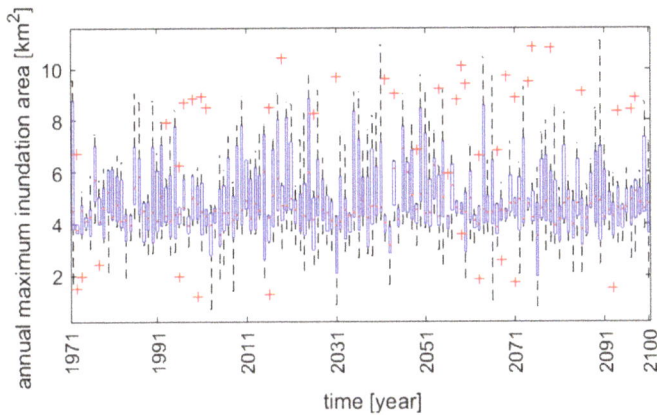

Figure 11. Maximum annual inundation from Models 1–7. Median: red lines in the boxes. The bottom and top edges of the blue box indicate the 25th and 75th percentiles, respectively. Maximum and minimum values: dashed black line in each box; red '+': outliers.

The projected maximum annual inundation areas for the seven models show a stationary character, which results from a stationary character of the annual maximum flow projections for the Biala Tarnowska [32]. However, looking at the outliers denoted by red crosses, the increase in annual maximum inundation areas can be found. Closer examination of precipitation projections (not shown here) reveals that 130-year daily raw precipitation data contain peaks of extreme precipitation, which are caused by the regional climate model instability (the so-called one-grid storms). These abnormal events may lead to wrong conclusions on the future projections of extreme high flow events when linear regression models are used to analyze future trends and the uncertainty of the projections is not taken into account [32].

The instabilities were detected by a comparison with the precipitation projections for the neighboring catchment Dunajec. The extreme event should appear in both catchments; otherwise, the extreme event is classified as a one-grid storm and replaced by the second-highest event in the year. In this study, instabilities in all but two models were detected. In the case of Biala Tarnowska, Models 2 and 5 were free of one-grid storms (Figure 12). Significant changes in maximum inundation return periods were found for Models 1, 3, 4, 6 and 7.

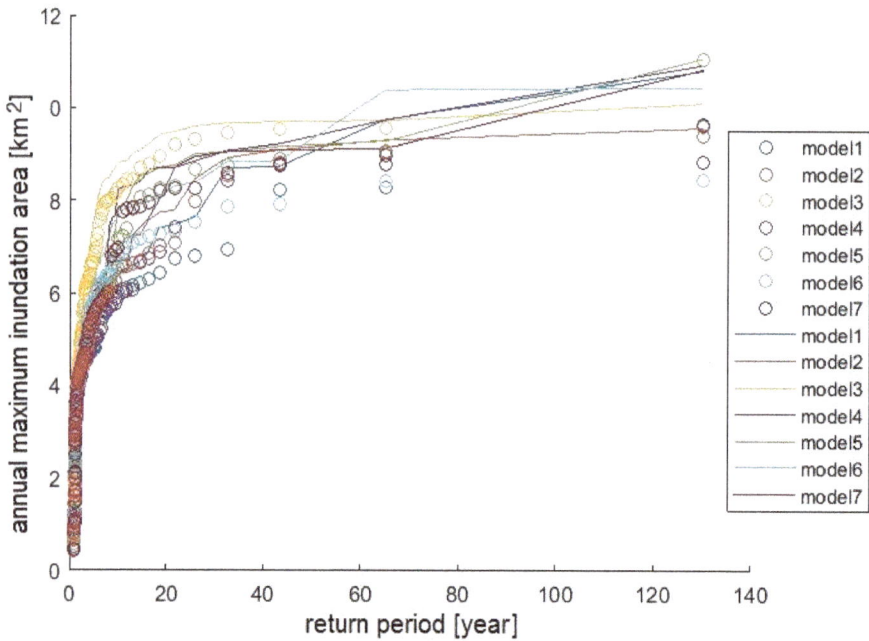

Figure 12. Return period for 130 annual maxima of inundation area for seven models without a one-grid-storm ('o') and with a one-grid-storm (solid line).

The NEVA routine [43] for the generalized extreme value (GEV) analysis was performed for maximum annual inundation area data presented in Figure 12. The one-grid storms influence the bounds of inundation projection uncertainty (Figure 13). The left panel presents an analysis of data without one-grid storms, while the right panel presents results with one-grid storms. Obviously, the uncertainty bounds are smaller in the case of data without RCM instabilities.

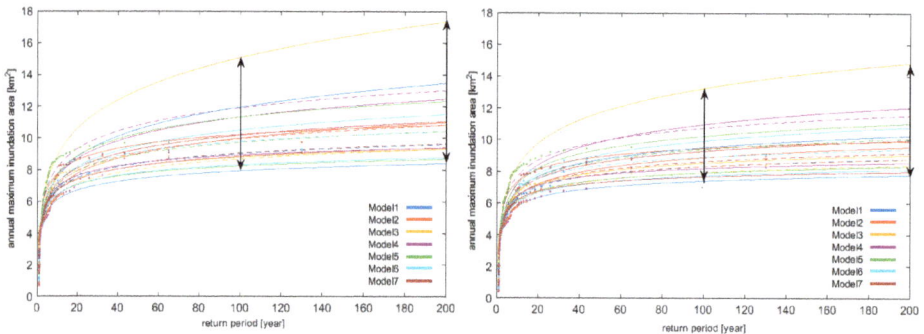

Figure 13. Annual maximum inundation area over a 100-year return period (marked with a black double arrow) and a 200-year return period (marked with a black double arrow), with instabilities (**left plot**) and without instabilities (**right plot**), the 95% and 5% confidence bounds (respectively with the Models, solid colorful lines), and MLE (dashed lines).

Removing one-grid storm events significantly change the results by decreasing the bounds of uncertainty of the inundation area.

4. Conclusions

The main issue related to the simulation approach within future climate change projections is the simulation time. The size, and above all the amount, of data derived from the projections that have to be processed by a distributed model make flood risk mapping very time-consuming. To accelerate the process, an emulator of a distributed model was applied. The emulator of the 1D MIKE11 was based on a stochastic transfer function that presents linear process dynamics, while the linearization of a flow–water-level relationship was performed with the Box–Cox transformation, which acts as an inverse of a rating curve. The use of the emulator significantly speeds up the process but may cause other problems, related mainly to simplification errors and the difficulty in assessing the uncertainty of its predictions when spatial observations are missing. However, the latter applies also to a distributed flow routing model. The way forward is further work on the development of remote sensing methods, using the unmanned aerial system (UAS) to derive local aerial photographs of the inundation extent.

Using the Tuchow meander as an example, relationships have been derived between the simulated water depth at a given cross section on the meander and the meander inundation area for the annual maximum flows (Figure 6). Those relationships can be used to assess the sensitivity of maximum inundation area to changes in annual maximum flows governed by climatic changes.

An additional advantage of the simulation approach presented here is the taking into account of the variability of antecedent conditions in the catchment before a flood event. The other approach to the derivation of flood risk maps that uses a design storm of a specified return period derived from the precipitation projections [44] results in inundation area projections that are biased and depend on assumed soil moisture conditions before the storm.

There is a large uncertainty in climate projections that influences the estimates of future flood risk. The crucial action in the case of adaptation to floods is a double check of climate projections. We show that the pre-processing of climate projections, by excluding one-grid storms (instabilities), significantly reduces the uncertainty of inundation projections.

Author Contributions: Conceptualization: J.D. and R.J.R.; methodology: J.D. and R.J.R.; software: J.D. and R.J.R. and A.K.; validation: R.J.R.; formal analysis: J.D.; investigation: J.D. and R.J.R.; data curation: J.D.; writing—original draft preparation: J.D. and R.J.R.; writing—review & editing: J.D., R.J.R., and A.K.; visualization: J.D. and R.J.R.; supervision: R.J.R.

Funding: This work was partially supported within statutory activities No 3841/E-41/S/2017 of the Ministry of Science and Higher Education of Poland.

Conflicts of Interest: The authors declare no conflict of interest.

References

1. European Flood Directive 2007. Available online: http://ec.europa.eu/environment/water/flood_risk/index.htm (accessed on 27 December 2018).
2. Popielarz-Deja, S.; Lesniak, E. Examples of Hydrological Computation for Water-Drainage assessments: Methodology and examples of hydrological evaluations in the case of lack of hydrological observations. In *Materialy Pomocnicze—Centralne Biuro Studiow i Projektow Wodnych Melioracji*; Centralne Biuro Studiów i Projektów Wodnych Melioracji: Warsaw, Poland, 1971. (In Polish)
3. Faulkner, D.; Wass, P. Flood estimation by continuous simulation in the Don catchment, South Yorkshire, UK. Water and Environment. *J. CIWEM* **2007**, *19*, 78–84.
4. Falter, D.; Schröter, K.; Dung, NV.; Vorogushyn, S.; Kreibich, H.; Hundecha, Y.; Apel, H.; Merz, B. Spatially coherent flood risk assessment based on long-term continuous simulation with a coupled model chain. *J. Hydrol.* **2015**, *524*, 182–193. [CrossRef]
5. Intergovernmental Panel on Climate Change (IPCC). *Climate Change 2014: Impacts, Adaptation, and Vulnerability. Part A: Global and Sectoral Aspects. Contribution of Working Group II to the Fifth Assessment Report of the Intergovernmental Panel on Climate Change*; IPCC: Geneva, Switzerland, 2014.
6. European Environment Agency (EEA). *Annual Report 2013 and Environmental Statement 2014*; Technical Report; European Environment Agency: Copenhagen, Denmark, 2013.

7. Tanaka, T.; Tachikawa, Y.; Ichikawa, Y.; Yorozu, K. Impact assessment of upstream flooding on extreme flood frequency analysis by incorporating a flood-inundation model for flood risk assessment. *J. Hydrol.* **2017**, *554*, 370–382. [CrossRef]

8. Doroszkiewicz, J.; Romanowicz, R.J. Guidelines for the adaptation to floods in changing climate. *Acta Geophys.* **2017**, *65*, 849–861. [CrossRef]

9. Apel, H.; Thieken, A.H.; Merz, B.; Blöschl, G. Flood risk assessment and associated uncertainty. *Nat. Hazards Earth Syst. Sci.* **2004**, *4*, 295–308. [CrossRef]

10. Komatina, D.; Branisavljevic, N. *Uncertainty Analysis as a Complement to Flood Risk Assessment*; Faculty of Civil Engineering University of Belgrade: Beograd, Serbia, 2005. Available online: http://daad.wb.tu-harburg.de/fileadmin/BackUsersResources/Risk/Dejan/UncertaintyAnalysis.pdf (accessed on 25 March 2016).

11. Kiczko, A.; Romanowicz, R.J.; Osuch, M.; Karamuz, E. Maximising the usefulness of flood risk assessment for the River Vistula in Warsaw. *Nat. Hazards Earth Syst. Sci.* **2013**, *13*, 3443–3455. [CrossRef]

12. Romanowicz, R.J.; Kiczko, A. An event simulation approach to the assessment of flood level frequencies: Risk maps for the Warsaw reach of the river Vistula. *Hydrol. Process.* **2016**, *30*, 2451–2462. [CrossRef]

13. Castelletti, A.; Galelli, S.; Restelli, M.; Soncini-Sessa, R. Data-driven dynamic emulation modelling for the optimal management of environmental systems. *Environ. Model. Softw.* **2012**, *34*, 30–43. [CrossRef]

14. Merz, B.; Thieken, A.H. Flood risk curves and uncertainty bounds. *Nat. Hazards* **2009**, *51*, 437–458. [CrossRef]

15. Water Act 20 July 2017 r. in Polish, Ustawa z dnia 20 lipca 2017 r. Prawo wodne (Dz.U. z 2017 poz. 1566). Available online: http://prawo.sejm.gov.pl/isap.nsf/DocDetails.xsp?id=WDU20170001566 (accessed on 22 February 2018).

16. Jones, R.N. Managing uncertainty in climate change projections—Issues for impacts assessments. *Clim. Chang.* **2000**, *45*, 403–419. [CrossRef]

17. Wilby, R.L.; Suraje, D. Robust adaptation to climate change. *Weather* **2010**, *65*, 180–185. [CrossRef]

18. Chen, I.-C.; Hill, J.K.; Ohlemüller, R.; Roy, D.B.; Thomas, C.D. Rapid range shifts of species associated with high levels of climate warming. *Science* **2011**, *333*, 1024–1026. [CrossRef] [PubMed]

19. Willems, P.; Vrac, M. Statistical precipitation downscaling for small-scale hydrological impact investigations of climate change. *J. Hydrol.* **2011**, *402*, 193–205. [CrossRef]

20. Piani, C.; Weedon, G.P.; Best, M.; Gomes, S.M.; Viterbo, P.; Hagemann, S.; Haerter, J.O. Statistical bias correction of global simulated daily precipitation and temperature for the application of hydrological models. *J. Hydrol.* **2010**, *395*, 199–215. [CrossRef]

21. Osuch, M.; Romanowicz, R.J.; Lawrence, D.; Wong, W.K. Trends in projections of standardized precipitation indices in a future climate in Poland. *Hydrol. Earth Syst. Sci.* **2016**, *20*, 1947–1969. [CrossRef]

22. Romanowicz, R.J.; Kiczko, A.; Napiorkowski, J.J. Stochastic transfer function model applied to combined reservoir management and flow routing. *Hydrol. Sci. J.* **2010**, *55*, 27–40. [CrossRef]

23. Foglia, L.; Hill, M.C.; Mehl, S.W.; Burlando, P. Sensitivity analysis, calibration, and testing of a distributed hydrological model using error-based weighting and one objective function. *Water Resour. Res.* **2009**, *45*. [CrossRef]

24. Romanowicz, R.J.; Bogdanowicz, E.; Debele, S.; Doroszkiewicz, J.; Hisdal, H.; Lawrence, D.; Meresa, H.; Napiorkowski, J.; Osuch, M.; Strupczewski, W.G.; et al. Climate change impact on hydrological extremes: Preliminary results from the Polish-Norwegian project. *Acta Geophys.* **2016**, *64*, 477–509. [CrossRef]

25. Bergstrom, S. The HBV model. In *Computer Models in Watershed Hydrology*; Singh, V.P., Ed.; Water Resources Publications: Highland Ranch, CO, USA, 1995; pp. 443–476.

26. Hamon, W.R. Estimating Potential Evapotranspiration. Bachelor's Thesis, Department of Civil and Sanitary Engineering, Massachusetts Institute of Technology, Cambridge, MA, USA, 1960.

27. Abbott, M.B.; Bathurst, J.C.; Cunge, J.A.; OConnell, P.E.; Rasmussen, J. An introduction to the European Hydrological System Systeme Hydrologique Europeen, SHE, 1: History and philosophy of a physically-based, distributed modelling system. *J. Hydrol.* **1986**, *87*, 45–59. [CrossRef]

28. Stelling, G.S.; Duinmeijer, S.P.A. A staggered conservative scheme for every Froude number in rapidly varied shallow water flows. *Int. J. Numer. Methods Fluids* **2003**, *43*, 1329–1354. [CrossRef]

29. Barkau, R. *UNET, One-Dimensional Flow through a Full Network of Open Channels*; User's Manual Version 2.1; Publication CPD-66; Hydrologic Engineering Center, US Army Corps of Engineers: Davis, CA, USA, 1996.

30. Cholet, C.; Charlier, J.-B.; Moussa, R.; Steinmann, M.; Denimal, S. Assessing lateral flows and solute transport during floods in a conduit-flow-dominated karst system using the inverse problem for the advectiondiffusion equation. *Hydrol. Earth Syst. Sci.* **2017**, *21*, 3635–3653. [CrossRef]

31. Storn, R.; Price, K.V. *Differential Evolution—A Simple and Efficient Adaptive Scheme for Global Optimization Over Continuous Spaces*; Technical Report TR-95-012; International Computer Sciences Institute: Berkeley, CA, USA, 1995.

32. Meresa, H.K.; Romanowicz, R.J. The critical role of uncertainty in projections of hydrological extremes. *Hydrol. Earth Syst. Sci.* **2017**, *21*, 4245–4258. [CrossRef]

33. Young, P.C. *Recursive Estimation and Time Series Analysis*; Springer-Verlag: Berlin, Germany, 1984.

34. Romanowicz, R.J.; Young, P.C.; Beven, K.J.; Pappenberger, F. A data based mechanistic approach to nonlinear flood routing and adaptive flood level forecasting. *Adv. Water Resour.* **2008**, *31*, 1048–1056. [CrossRef]

35. Nelder, J.A.; Mead, R. A Simplex Method for Function Minimization. *Comput. J.* **1965**, *7*, 308–313. [CrossRef]

36. Romanowicz, R.J.; Doroszkiewicz, J.M. The Application of Cumulants to Flow Routing. *Meteorol. Hydrol. Water Manag.* **2018**, *7*. [CrossRef]

37. Aronica, G.; Bates, P.D.; Horritt, M.S. Assessing the uncertainty in distributed model predictions using observed binary pattern information within glue. *Hydrol. Process.* **2002**, *16*, 2001–2016. [CrossRef]

38. Horritt, M.S.; Bates, P.D. Predicting floodplain inundation: Raster-based modelling versus the finite-element approach. *Hydrol. Process.* **2001**, *15*, 825–842. [CrossRef]

39. Romanowicz, R.J.; Osuch, M. Assessment of land use and water management induced changes in flow regime of the Upper Narew. *Phys. Chem. Earth* **2011**, *36*, 662–672. [CrossRef]

40. Steenbergen, N.; Willems, P. Method for testing the accuracy of rainfall–runoff models in predicting peak flow changes due to rainfall changes, in a climate changing context. *J. Hydrol.* **2012**, *414–415*, 425–434. [CrossRef]

41. Kotlarski, S.; Keuler, K.; Christensen, O.B.; Colette, A.; Déqué, M.; Gobiet, A.; Goergen, K.; Jacob, D.; Lüthi, D.; van Meijgaard, E.; et al. Regional climate modeling on European scales: A joint standard evaluation of the EURO-CORDEX RCM ensemble. *Geosci. Model Dev.* **2014**, *7*, 1297–1333. [CrossRef]

42. Teutschbein, C.; Seibert, J. Bias correction of regional climate model simulations for hydrological climate-change impact studies: Review and evaluation of different methods. *J. Hydrol.* **2012**, *456–457*, 12–29. [CrossRef]

43. Cheng, L.; AghaKouchak, A.; Gilleland, E.; Katz, R.W. Non-stationary extreme value analysis in a changing climate, climatic change. *Clim. Chang.* **2014**, *127*, 353–369. [CrossRef]

44. Papaioannou, G.; Efstratiadis, A.; Vasiliades, L.; Loukas, A.; Papalexiou, S.; Koukouvinos, A.; Tsoukalas, I.; Kossieris, P. An operational method for flood directive implementation in ungauged urban areas. *Hydrology* **2018**, *5*, 24. [CrossRef]

water

MDPI

Article

Flooding Related Consequences of Climate Change on Canadian Cities and Flow Regulation Infrastructure

Ayushi Gaur [1], Abhishek Gaur [2,*], Dai Yamazaki [3] and Slobodan P. Simonovic [1]

[1] Facility for Intelligent Decision Support, Department of Civil and Environmental Engineering, Western University, London, ON N6A 3K7, Canada; ayushigaur.evs@gmail.com (A.G.); simonovic@uwo.ca (S.P.S.)
[2] National Research Council Canada, 1200 Montreal Road, Ottawa, ON K1A 0R6, Canada
[3] Institute of Industrial Science, The University of Tokyo, Meguro-ku, Tokyo 153-8505, Japan; yamadai@rainbow.iis.u-tokyo.ac.jp
* Correspondence: Abhishek.Gaur@nrc-cnrc.gc.ca; Tel.: +1-613-998 9799

Received: 26 October 2018; Accepted: 27 December 2018; Published: 1 January 2019

Abstract: This study discusses the flooding related consequences of climate change on most populous Canadian cities and flow regulation infrastructure (FRI). The discussion is based on the aggregated results of historical and projected future flooding frequencies and flood timing as generated by Canada-wide hydrodynamic modelling in a previous study. Impact assessment on 100 most populous Canadian cities indicate that future flooding frequencies in some of the most populous cities such as Toronto and Montreal can be expected to increase from 100 (250) years to 15 (22) years by the end of the 21st century making these cities highest at risk to projected changes in flooding frequencies as a consequence of climate change. Overall 40–60% of the analyzed cities are found to be associated with future increases in flooding frequencies and associated increases in flood hazard and flood risk. The flooding related impacts of climate change on 1072 FRIs located across Canada are assessed both in terms of projected changes in future flooding frequencies and changes in flood timings. Results suggest that 40–50% of the FRIs especially those located in southern Ontario, western coastal regions, and northern regions of Canada can be expected to experience future increases in flooding frequencies. FRIs located in many of these regions are also projected to experience future changes in flood timing underlining that operating rules for those FRIs may need to be reassessed to make them resilient to changing climate.

Keywords: climate change; flood hazard; flood risk; return period; streamflow regulation rules; Canada

1. Introduction

Due to consistent warming of Earth's climate particularly in the recent decades, weather elements and their extremes have become more severe and frequent, with increases in extreme heat, intense precipitation, drought, and flooding recorded across the globe [1,2]. Floods are the most frequently occurring natural hazard in Canada and are shaped by one or more hydrologic processes such as snowmelt and ice jams, hydro-meteorological factors such as short duration intense rainfall, rain on snow, or structural failures such as dam breaks [3,4].

Evidences of shifts in extreme flow characteristics have been recorded in different regions of Canada. Trends in the flow extremes for 248 catchments in Canada were examined by Burn et al. [5]. Decreasing trends in annual maximum flows were obtained for catchments located in southern Canada and increasing trends were obtained for catchments located in northern Canada. Springtime and summer runoff characteristics were examined by Burn et al. [6] for stations located in the Canadian prairies and obtained decreasing trends in spring snowmelt runoff volume and peak flows, and an earlier occurrence of the flooding events. Changes in the peak flow timings among rivers located in the western Canada over the time-period: 1960–2006 were analyzed by Dery et al. [7]. An earlier onset of spring melt, decrease in summer streamflow, and delay in the onset of enhanced autumn flows were obtained in most of the analyzed rivers. Flow trends at 68 gauging stations with at least 50 years of flow records were analyzed by Burn et al. [8], and more trends than to be expected by chance were found. They noted decreasing trends in annual and spring maximum flows and an earlier shift in the timing of the flooding events. Increasing magnitudes and frequencies of extreme floods has affected Canadian cities and population therein. Indeed, severe flooding events have been recorded in some of the most populous Canadian cities in the recent years such as: Montreal in 2017 and 2012, Thunder Bay in 2012, Calgary in 2013 and 2010, Winnipeg in 2009, and Toronto in 2005 and 2013 [9]. The enormous costs of Calgary floods in 2013 (over $5 billion) and Toronto floods in 2005 ($587 million) and 2013 ($1.2 billion) provide some examples of costs associated with urban flooding in Canada.

A number of approaches have been used to simulate flood magnitudes and assess flooding frequencies in catchments across the globe. Gaur et al. [10] assessed projected future changes in flood hazard for the Grand River basin, Canada by generating flows at the catchment outlet using WatFlood hydrologic model and by fitting a Generalized Pareto distribution (GPD) on the peak-over-threshold values of simulated flooding events. Fiorentino et al. [11] and Gioia et al. [12] demonstrated the use of spatially discrete parameters in theoretically derived distributions and analytical models to more accurately perform regional flood frequency analysis in Italy. Devkota and Gyawali [13] assessed the impacts of climate change on the Koshi River basin in Nepal by using bias-corrected climate forecasts from two Regional Climate Models (RCMs) and performing hydrological simulations using the SWAT model. Qin and Lu [14] assessed projected future changes in flooding frequencies for the Heshui watershed in China. A coupled Long Ashton Research Station Weather Generator (LARS-WG) and Semi-distributed Land Use–Based Runoff Processes (SLURP) approach was used to generate future flows, which were then fitted to Pearson type III distribution to perform flood frequency analysis. Das et al. [15] evaluated the suitability of three extreme value distributions: Gumbel, Log-Pearson type 3, and Generalized Extreme Value (GEV) in modelling flood extremes and recommended the use of GEV for climate change impact assessments.

Long-term forecasts of flood-hazard and flood-risk have been made at national and continental scales have been made [16–22]. For instance, future changes in flood hazard across Europe was assessed by dynamically downscaling future climatic projections from two regional climate models (RCMs) and performing hydrologic modelling using LISFLOOD model [23]. Streamflow was generated for historical (1961–1990) and future (2061–2100) timelines and an assessment of projected changes in flood magnitudes of 100 year return period floods was performed. It was estimated that the recurrence interval of the present day 100-year flood event might decrease to 50 years or less in many parts of Europe in the future. Yamazaki et al. [22] used coarse resolution runoff simulations from the GCMs and simulated higher resolution water levels across the Amazon River basin using CaMa-Flood hydrodynamic model. Hirabayashi et al. [20] assessed future changes in global flood hazard under projected climate change influences using runoff projections from 11 Global Climate Models (GCMs) and CaMa-Flood hydrodynamic model [24]. In Gaur et al. [16] a similar assessment to Hirabayashi et al. [20] was performed for the Canadian landmass but taking into consideration a larger ensemble of runoff projections made by 21 CMIP5 GCMs.

While information on the distribution of flood hazard at a national to global scales is extremely valuable, knowledge about how it is distributed with reference to the population, and built assets is of critical importance to water resource managers, city planners, and policy-makers. Typically flood hazard is combined with an exposure of interest to calculate flood risk. Several techniques have been used in the past to assess flood risk at national to global scales. Hirabayashi et al. [20] for instance quantified global flood exposure by combining gridded population information with the gridded flooded area modelled using the CaMa-Flood model. Jongman et al. [25] demonstrated two methods to assess damages caused by flooding at a global scale. In the first method, the economic damage caused by floods for a particular country was calculated by combining information on the population exposed to flooding and GDP per capita of the country whereas in the second method, it was calculated by finding the total urban area exposed to flooding and combining it with the information on maximum damage per square meter for the country. Peduzzi et al. [26] evaluated a Disaster Risk Index by combining the hazards such as floods, droughts, cyclones, and earthquakes with gridded population to assess the human exposure to these hazards at a global scale. De Moel et al. [27] evaluated historical and projected future changes in flood exposure by overlaying information on the distribution of historical and projected future flood depths with the distribution of the urban agglomerations in the Netherlands. Kleinen and Petschel-Held [28] summed the population living in river basins across the globe where the return period of historical 50-year return period event is projected to reduce as a consequence of climate change. Feyen et al. [29,30] combined flood frequency curves with flood depth-damage functions to estimate current and future average annual damage in Europe.

The objective of this study is to use the historical and projected future flood characteristics simulated in Gaur et al. [16], and quantify the risk the projected changes pose to the Canadian population, and the network of flow regulation infrastructure (FRI) existing in Canada. A novel methodology to quantify and demonstrate this risk has been presented in this study using which cities and FRIs most at risk from projected changes in flooding frequencies and timings have been identified. To the best of our knowledge, an assessment of this scale has not been presented before for the Canadian cities and FRIs highlighting the novelty of both the adopted methodology and generated results. The discussion presented in this study will help water resource managers and policymakers to formulate flood management guidelines in Canada.

The rest of the paper is organized as follows. Description of the methodology used for estimating changes in flood hazard and flood risk at Canadian cities and regulated flow gauging stations is provided in Section 2. This is followed by a description of the study region and datasets used in Section 3. Results and discussion are provided in Section 4, followed by conclusions in Section 5.

2. Methodology

In Gaur et al. [16] future flow projections were made for the entire Canadian landmass at ~25 km spatial resolution by adopting an approach used in previous studies at other locations [20,22]. A large ensemble of runoff projections from 21 GCMs were used to obtain flood characteristics across Canada under four different Representative Concentration Pathways (RCPs): RCP 2.6, RCP 4.5, RCP 6.0, and RCP 8.5 [31] for historical (1961–2005) and future (2061–2100) timelines. The projected future changes in flooding frequencies of historical 100 year and 250 year return period events were calculated along with the changes in the timing of the floods. The results were discussed in terms of the aggregated projected changes and associated uncertainty contributed by different GCMs.

The aggregation of the projected changes for a particular RCP was performed using two approaches: (1) *all GCM median approach* where median of the changes projected by all GCMs was considered as the aggregated change, and (2) *robust GCM median approach* where median of the projected changes made only by the robust GCMs was considered as the aggregated change. When applying the *robust GCM median approach* to aggregate projected flooding frequency changes, robust projections were considered as the ones whose projected sign of change was concurred upon by more than 50% (or the majority) of the GCMs. In the case of flood timing, the robust projections were considered as the ones whose month of flood timing was concurred upon by more than 50% of the GCMs considered. More details on the methodology used to prepare the aggregated projections can be obtained from Gaur et al. [16].

In this study, the aggregated results from Gaur et al. [16] are used to investigate the flooding related consequences of climate change on 100 most populous Canadian cities and 1072 Flow Regulation Infrastructure (FRI) located in Canada.

2.1. CaMa-Flood Hydrodynamic Model

CaMa-Flood [22,24,32] is a global scale distributed hydrodynamic model that routes input runoff generated by a land surface model to oceans or inland seas along a prescribed river network map. Water storage is calculated at every time-step, whereas variables such as: water level, inundated area, river discharge, and flow velocity, are diagnosed from the calculated water storage. River discharge and flow velocity are estimated using local inertial equation. Floodplain inundation is modelled by taking into consideration sub-grid scale variabilities in river channel and floodplain topography. A river channel reservoir has three parameters: channel length (L), channel width (W), and bank height (B). The floodplain reservoir has a parameter for unit catchment area (A_c), and a floodplain elevation profile which describes floodplain water depth D_f as a function of flooded area A_f. The topography related parameters i.e., surface altitude (Z), distance to downstream cell (X), and unit catchment area (A_c) that are calculated using the Flexible Location of Waterways (FLOW) method [32]. Finally, a Manning's roughness coefficient parameter (n) is used to represent the roughness in the river channel.

The CaMa-Flood model has been validated extensively for its ability to simulate runoff in some of the largest catchments of the globe including the Amazon, Mississippi, Parana, Niger, Congo, Ob, Ganges, Lena, and Mekong [22,24,33]. Given the high credibility of the model in simulating river flow and flood inundation dynamics, the model has been used to assess the impacts of climate change at regional to global scales [19,20,33–36]. This study uses the globally calibrated version of Cama-Flood model that was used to generate global scale runoff projections in Hirabayashi et al. [20].

2.2. Future Projected Changes in Flood Risk in Canadian Cities

The projected future changes in flood risk are calculated for the Canadian cities by combining the projected future changes in flood hazard and population exposed to flooding at the selected cities. Flood hazard is expressed as a function of the magnitude of the projected changes in the flooding frequencies at the city locations. The population exposed to flooding is calculated from flooded area estimates obtained from the results of a retrospective CaMa-Flood simulation.

2.2.1. Projected Changes in Flood Hazard

In this study, the projected future changes in the frequencies of historical 100-year and 250-year flooding events are used to define future changes in flood hazard for the cities. Future return periods of historical 100-year and 250-year return period flows were obtained in Gaur et al. [20] by first estimating the historical flood magnitudes corresponding to 100-year and 250-year return periods. This is done by fitting a Generalized Extreme Value (GEV) distribution on annual maximum historical flows ($GEV_{his,f}$). Next, a GEV distribution is fitted on the annual maximum projected future flows ($GEV_{fut,f}$). Thereafter, future return periods of historical 100-year and 250-year return period flows are estimated by using applying the inverse of $GEV_{fut,f}$ function on historical 100-year and 250-year flow magnitudes. The aggregated results from both *all GCM median* and *robust GCM median* methods obtained in Gaur et al. [20] are extracted for grids encompassing the selected city domains, and used to calculate projected changes in flood hazard using Equation (1).

$$\Delta FH_c = \frac{(RP_{c,h} - RP_{c,f})}{\max(RP_{c,f}) - \min(RP_{c,f})} \tag{1}$$

where ΔFH_c denotes the flood hazard index that depends on the values of projected changes in flood hazard projected between historical (h) and future (f) timelines at city c, RP denotes the return-period of the floods analyzed, and max (min) represent the maximum (minimum) values of return periods among the selected cities.

2.2.2. Population Exposed to Flooding

The total population exposed to flooding is calculated by performing a retrospective simulation of CaMa-Flood following previous studies [20,25]. Following steps are performed:

- A retrospective simulation of a land surface model MATSIRO [37] forced by climate variables obtained from gauges and reanalysis datasets, and with CaMa-Flood river routing is performed for the time-period 1979–2010 for the entire Canadian landmass at 0.005° spatial resolution to simulate daily discharges, water levels and other flood inundation related variables. The estimated discharge and hydrologic variables from this retrospective run has been validated against observations from Gravity Recovery and Climate Experiment (GRACE) based terrestrial water storage data in Kim et al. [38].
- The annual maximum discharge and water levels obtained from this retrospective run are fitted to a Generalized Extreme Value (GEV) distribution and their 100 and 250 year return period estimates are obtained.
- The water levels corresponding to 100 and 250 year return period flooding events are used to obtain the fraction of CaMa-Flood grids encompassing different cities that are flooded as a consequence of these events.
- The flooded area fraction is then multiplied to the total population of the cities to obtain the total population exposed to flooding.

2.2.3. Projected Changes in Flood Risk

The normalized aggregated changes in flood-hazard ΔFH_c projected for the cities are multiplied with the normalized log of population exposed to flooding (P_f) to calculate future projected changes in flooding risk in terms of flood risk index ΔFR_c as described in Equation (2).

$$\Delta FR_c = \Delta FH_c \times \frac{\ln P_f - \min(\ln P_f)}{\max(\ln P_f) - \min(\ln P_f)} \tag{2}$$

2.3. Future Changes in Flood Hazard at Flow Regulation Infrastructure Locations in Canada

The Flow Regulation Infrastructure (FRI) are sensitive to both the magnitudes and timings of flooding events. Therefore, projected changes in both these aspects of flooding are discussed to highlight changing flood hazard at FRIs located in Canada.

3. Study Region and Data Used

Canada is the second largest country in the world with a total land mass of ~10 million square kilometers. It is part of North America and consists of ten provinces and three territories: Yukon (YK), Northwest Territories (NT), Nunavut (NV), British Columbia (BC), Alberta (AB), Saskatchewan (SK), Manitoba (MB), Ontario (ON), Quebec (QB), Newfoundland and Labrador (NL), New Brunswick (NB), Nova Scotia (NS), and Prince Edward Island (PEI), as shown in Figure 1.

The list of 100 most populous cities and their population for the year 2015 is obtained from Statistics Canada (https://www.statcan.gc.ca/eng/start). The location of flow regulation infrastructure is obtained from the HYDAT flow database maintained by the Environment and Climate Change Canada (EC Data Explorer 2016). In the HYDAT database, 1072 FRIs have been identified. The distribution of these locations across Canada is shown in Figure 1.

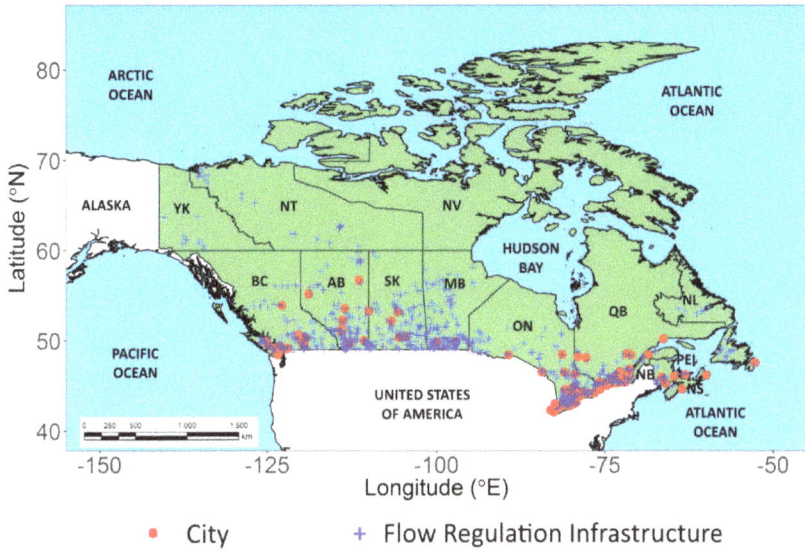

Figure 1. Canada, its administrative boundaries, surrounding countries, and waterbodies. The location of cities and flow regulation locations considered for assessment in this study are shown as red dots and blue crosses respectively.

4. Results and Discussion

4.1. Projected Changes in Flood-Risk in Canadian Cities

The sign of change in flooding frequencies obtained for 100 most populated cities in Canada obtained by all GCM median and robust GCM median approaches is summarized in Table 1 for 100-year and 250-year return period flooding events. It can be noted that all RCPs and return periods considered, 40–60% of the cities are projected with future increases in flooding frequencies. Largest numbers of cities with increasing flooding frequencies are projected for RCP 8.5, followed by RCP 6.0, and then followed by RCP 2.6 and RCP 4.5. In addition, the total numbers of cities projected with future increases in flooding frequencies is found to be fractionally larger for 250-year flooding event than 100-year flooding event. Only a small difference in the total numbers of cities projected with future increases in flooding frequencies is obtained between the two methods of aggregation.

Table 1. Sign of change in flooding frequencies projected for 100 most populous Canadian cities.

RCP	100-Year			250-Year		
	Decrease	**Increase**	**No Change**	**Decrease**	**Increase**	**No Change**
	All GCM Median					
RCP 2.6	50	49	1	52	47	1
RCP 4.5	55	43	2	51	48	1
RCP 6.0	46	53	1	34	65	1
RCP 8.5	34	60	6	32	63	5
	Robust GCM median					
RCP 2.6	50	49	1	52	47	1
RCP 4.5	56	43	1	51	48	1
RCP 6.0	46	53	1	35	64	1
RCP 8.5	39	60	1	35	63	1

A summary of future projected flooding frequencies for ten most populated Canadian cities obtained from robust GCM median and all GCM median approaches is summarized in Tables 2 and 3 respectively. It can be noted that some of most populous Canadian cities such as Toronto and Montreal are projected with future increases in flooding frequencies whereas other major cities such as Ottawa and Edmonton have been projected with future decreases in the flooding frequencies in the future. Other major cities such as Vancouver, Calgary, and Quebec have been projected with a mix of future increases and decreases in flooding frequencies for different sets of emission scenarios.

Table 2. Projected future return periods of historical 100-year and 250-year return period flood events for 10 most populated Canadian cities obtained from robust GCM median approach. Return periods greater than 200 (500) are only provided categorically in case of 100-year (250-year) return period events. Cases projected with future increases in flooding frequencies are shown in blue whereas cases projected with future decreases in flooding frequencies are shown in orange.

City	RCP 2.6		RCP 4.5		RCP 6.0		RCP 8.5	
	100-Year	250-Year	100-Year	250-Year	100-Year	250-Year	100-Year	250-Year
Toronto	22	37	32	46	32	39	15	22
Montreal	26	38	22	32	18	25	11	16
Vancouver	≥200	≥500	≥200	≥500	20	28	≥200	32
Calgary	46	94	≥200	≥500	≥200	≥500	49	85
Ottawa	≥200	≥500	≥200	≥500	≥200	≥500	≥200	≥500
Edmonton	104	284	≥200	≥500	≥200	≥500	≥200	≥500
Hamilton	37	56	151	78	≥200	67	27	56
Quebec	≥200	≥500	≥200	≥500	27	49	26	≥500
Winnipeg	≥200	≥500	≥200	≥500	≥200	≥500	≥200	≥500
Kitchener	26	47	≥200	86	≥200	57	29	60

Table 3. Projected future return periods of historical 100-year and 250-year return period flood events for 10 most populated Canadian cities obtained from all GCM median approach. Return periods greater than 200 (500) are only provided categorically in case of 100-year (250-year) return period events. Cases projected with future increases in flooding frequencies are shown in blue whereas cases projected with future decreases in flooding frequencies are shown in orange.

City	RCP 2.6		RCP 4.5		RCP 6.0		RCP 8.5	
	100-Year	250-Year	100-Year	250-Year	100-Year	250-Year	100-Year	250-Year
Toronto	30	63	39	86	116	227	23	35
Montreal	47	95	24	58	48	178	15	22
Vancouver	≥200	≥500	193	≥500	34	62	107	157
Calgary	96	199	133	254	≥200	≥500	87	156
Ottawa	≥200	≥500	≥200	≥500	≥200	≥500	≥200	≥500
Edmonton	104	284	189	476	≥200	≥500	≥200	≥500
Hamilton	79	142	102	223	125	198	76	174
Quebec	175	≥500	≥200	≥500	40	98	72	250
Winnipeg	151	416	191	393	≥200	≥500	121	319
Kitchener	58	108	103	231	126	188	32	74

The historical and future flooding frequencies are used to calculate flood-hazard index for different cities. The magnitudes obtained from robust GCM approach are shown in Figure 2. Note that in the figure, flood-hazard indices smaller than −1 are only presented categorically to show the spatial heterogeneity of hazard across the cities. It is found that the cities located in the southern Ontario and western Canada are projected with increases in flood hazard index suggesting future increases in flood hazard in the cities located in these regions. Similar spatial distribution of the hazard magnitudes is also obtained from the results of all GCM median approach (not shown).

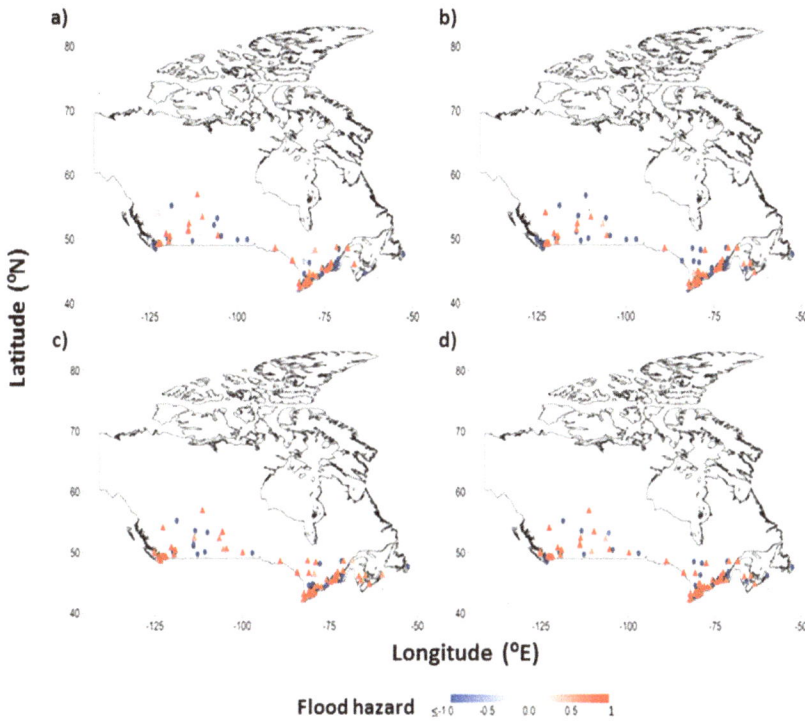

Figure 2. Flood hazard indices for 100 most populated cities in Canada under (**a**) RCP 2.6; (**b**) RCP 4.5; (**c**) RCP 6.0, and (**d**) RCP 8.5 obtained from robust GCM median approach. Positive values are presented as triangles whereas negative values are presented as dots. The index values below −1 are only shown categorically to show the spatial heterogeneity of the indices.

The calculated flood hazard index is combined with the population exposed to flooding to obtain flood risk index at each city. A scatterplot showing the projected changes in flood risk index values for historical 100-year return period flooding event (for cities projected with an increase in flood risk) and associated flood hazard index, and population, is presented in Figure 3. It can be seen that, as expected, both projected changes in flood hazard index and population exposed to flooding influence the calculated flood risk values at the cities.

Among the cities analyzed, the ten cities that have been projected with largest increases in flood-hazard and flood-risk as obtained from robust GCM median and all GCM median approaches are shown in Tables 4 and 5 respectively. It is noted that some of the cities associated with the highest flood hazard (such as Saint Catharines-Niagara, Nanticoke, Sault Ste. Marie) and risk (such as Saint Catharines-Niagara, Niagara, Toronto) are obtained from the results of both aggregation approaches however some differences in the rankings is also noted. It is particularly noted that the highest numbers of cities featuring in these tables are located in the Ontario province along with some highly populated cities such as Toronto, Montreal, and Saint Catharines-Niagara highlighting large increases in flood-hazard and flood-risk projected for the province. It is also noted that some smaller cities such as Walnut Grove (ON) are also associated with highest flood-hazard magnitudes but lower flood-risks due to smaller population exposed to flooding in these cities.

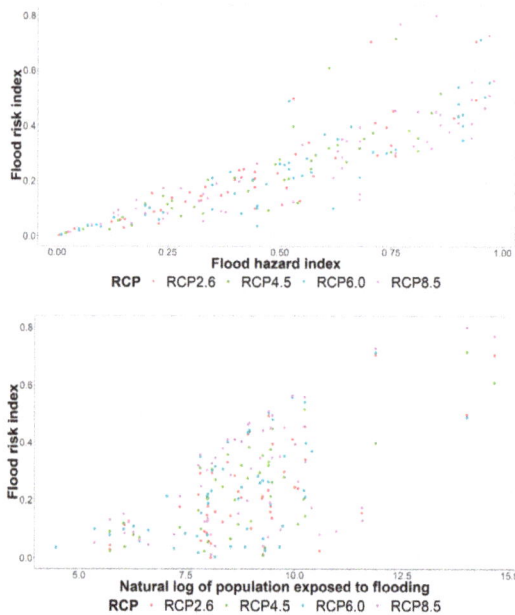

Figure 3. Scatterplot of future projected increases in flood risk with increases in projected flood hazard index, and calculated population exposed to flooding in the Canadian cities. The results are based on all GCM median approach.

4.2. Flooding Related Impacts on Flow Regulation Infrastructures (FRIs)

The flooding related impacts of climate change on FRIs are assessed by analyzing future projected changes in flood frequencies and month of extreme flows at their locations. The summary of sign of change in flooding frequencies obtained at the 1072 FRI locations is presented in Table 6. The spatial distribution of future flooding frequencies of historical 100 year return period floods is presented in Figures 4 and 5 for results obtained from robust GCM median and all GCM median aggregation approaches respectively. Results presented in Table 6 indicate that 40–50% of the FRIs located in Canada can be subjected to future increases in flooding frequencies under different sets of emission scenarios. From Figures 4 and 5 it can be noted that a vast majority of these FRIs are located in the south-western Ontario, west coast, and northern regions of Canada. A comparison of the results obtained from robust GCM median and all GCM median approaches indicate similar spatial distribution of the projected flooding frequencies among the FRIs but larger absolute magnitudes in the former than the latter. This can be noted both from the results presented in Figures 4 and 5 as well as Table 6 where as a consequence of this, a larger number of FRIs with no projected change in flooding frequencies obtained in the all GCM median case than the robust GCM median case. For instance, a considerable number (~10%) of the FRIs are projected with no change in flooding frequencies from all GCM median approach under RCP 8.5 which is considerably higher than the FRIs projected under robust GCM median approach (~1%). Between the four RCPs, the largest numbers of FRIs projected with future increases in flooding frequencies are obtained for RCP 6.0, followed by RCP 2.6, RCP 8.5 and RCP 4.5. The FRIs projected with the largest increases in flooding frequencies of historical 100 year return period flooding events from robust GCM median and all GCM median approaches are summarized in Tables 7 and 8 respectively. It can be noted from the results that the FRIs located in the prairies provinces such as Saskatchewan, Manitoba and Alberta, northern provinces such as Nunavut, and Ontario are found to be associated with the highest increases in future flooding frequencies from both aggregation methods.

Table 4. Ten cities projected with most increases in historical 100-year return period flood-hazard and flood-risk following robust GCM median approach. Information on the province and projected future flood frequency of historical 100-year return period floods in years is also provided within brackets.

S. No.	RCP 2.6		RCP 4.5		RCP 6.0		RCP 8.5	
	Hazard	Risk	Hazard	Risk	Hazard	Risk	Hazard	Risk
1	Sault Ste. Marie (ON; 4)	Toronto (ON; 22)	Saint Catharines-Niagara (ON; 4)	Montreal (QB; 22)	Halifax (NS; 1)	Montreal (QB; 18)	Sault Ste. Marie (ON; 2)	Toronto (ON; 15)
2	Nanticoke (ON; 5)	Saint Catharines-Niagara (ON; 6)	Nanticoke (ON; 4)	Toronto (ON; 32)	Sault Ste. Marie (ON; 3)	Toronto (ON; 32)	Saint Catharines-Niagara (ON; 3)	Montreal (QB; 11)
3	Saint Catharines-Niagara (ON; 6)	Montreal (QB; 26)	Cornwall (ON; 7)	Saint Catharines-Niagara (ON; 4)	Saint Catharines-Niagara (ON; 4)	Vancouver (BC; 20)	Cornwall (ON; 3)	Saint Catharines-Niagara (ON; 3)
4	Cornwall (ON; 7)	Sault Ste. Marie (ON; 7)	Salaberry-de-Valleyfield (QB; 7)	Abbotsford (BC; 11)	Nanticoke (ON; 4)	Saint Catharines-Niagara (ON; 4)	Nanticoke (ON; 3)	Quebec (QB; 26)
5	Salaberry-de-Valleyfield (QB; 7)	Nanticoke (ON; 5)	Walnut Grove (BC; 10)	Barrie (ON; 31)	Cornwall (ON; 6)	Halifax (NB; 1)	Salaberry-de-Valleyfield (QB; 4)	Hamilton (ON; 27)
6	Brantford (ON; 18)	Abbotsford (BC; 21)	Abbotsford (BC; 11)	Chicoutimi-Jonquiere (QB; 21)	Shawinigan (QB; 7)	Quebec (QB; 27)	White rock (BC; 5)	London (ON; 20)
7	Chatham (ON; 18)	Sarnia (ON; 21)	White Rock (BC; 11)	Thunder Bay (ON; 32)	Sorel (QB; 8)	Victoria (BC; 30)	Abbotsford (BC; 6)	Kitchener (ON; 29)
8	Saint-Jean-Sur-Richelieu (QB; 20)	Chatham (ON; 18)	Chilliwack (BC; 11)	Peterborough (ON; 17)	Joliette (QB; 8)	Abbotsford (BC; 9)	Chilliwack (BC; 6)	Vancouver (BC; 49)
9	Beloeil (QB; 20)	Hamilton (ON; 37)	Shawinigan (QB; 13)	White Rock (BC; 11)	Prince Albert (SK; 8)	Saskatoon (SK; 38)	Joliette (QB; 6)	Calgary (AB; 25)
10	Kamloops (BC; 20)	Cornwall (ON; 7)	Sorel (QB; 14)	Brantford (ON; 31)	White Rock (BC; 9)	Regina (SK; 40)	Walnut Grove (BC; 6)	Halifax (NB; 24)

Table 5. Ten cities projected with most increases in flood-hazard and flood-risk of historical 100-year return period floods in future obtained from all GCM median approach. Information on the province and projected future flood frequency magnitudes in years is also provided within brackets.

S. No.	RCP 2.6 Hazard	RCP 2.6 Risk	RCP 4.5 Hazard	RCP 4.5 Risk	RCP 6.0 Hazard	RCP 6.0 Risk	RCP 8.5 Hazard	RCP 8.5 Risk
1	Saint Catharines-Niagara (ON; 6)	Saint Catharines-Niagara (ON; 6)	Abbotsford (BC; 14)	Montreal (QB; 24)	Sault Ste. Marie (ON; 3)	Saint Catharines-Niagara (ON; 5)	Sault Ste. Marie (ON; 2)	Montreal (QB; 15)
2	Nanticoke (ON; 6)	Toronto (ON; 30)	Chilliwack (BC; 16)	Toronto (ON; 39)	Nanticoke (ON; 4)	Sault Ste. Marie (ON; 3)	Saint Catharines-Niagara (ON; 3)	Toronto (ON; 23)
3	Cornwall (ON; 10)	Montreal (QB; 47)	White Rock (BC; 19)	Abbotsford (BC; 14)	Saint Catharines-Niagara (ON; 5)	Abbotsford (BC; 10)	Nanticoke (ON; 3)	Saint Catharines-Niagara (ON; 3)
4	Salaberry-de-Valleyfield (QB; 11)	Nanticoke (ON; 6)	Walnut Grove (BC; 19)	Chilliwack (BC; 16)	White Rock (BC; 9)	Nanticoke (ON; 4)	Cornwall (ON; 4)	Sault Ste. Marie (ON; 2)
5	Walnut Grove (BC; 24)	Abbotsford (BC; 25)	Cornwall (ON; 21)	Saint Catharines-Niagara (ON; 47)	Cornwall (ON; 9)	Montreal (QB; 48)	Salaberry-de-Valleyfield (QB; 4)	Abbotsford (BC; 7)
6	Abbotsford (BC; 25)	Cornwall (ON; 10)	Montreal (QB; 24)	Cornwall (ON; 21)	Walnut Grove (BC; 9)	Chilliwack (BC; 10)	Abbotsford (BC; 7)	Nanticoke (ON; 3)
7	White Rock (BC; 25)	Salaberry-de-Valleyfield (QB; 11)	Kamloops (BC; 25)	Belleville (ON; 29)	Abbotsford (BC; 10)	Sorel (BC; 15)	White rock (BC; 7)	Chilliwack (BC; 7)
8	Chilliwack (BC; 26)	Sault Ste. Marie (ON; 29)	Belleville (ON; 29)	White Rock (BC; 19)	Chilliwack (BC; 10)	Cornwall (ON; 9)	Chilliwack (BC; 7)	Cornwall (ON; 4)
9	Kamloops (BC; 28)	Chilliwack (BC; 26)	Saint-jean-sur-richelieu (QB; 31)	Saint-jean-sur-richelieu (QB; 31)	Salaberry-de-Valleyfield (QB; 10)	Salaberry-de-Valleyfield (QB; 10)	Walnut Grove (BC; 7)	Salaberry-de-Valleyfield (QB; 4)
10	Sault Ste. Marie (ON; 29)	Thunder Bay (ON; 44)	Beloeil (QB; 32)	Nanticoke (ON; 37)	Joliette (QB; 10)	Joliette (QB; 10)	Joliette (QB; 10)	Sarnia (ON; 24)

Table 6. Sign of change in flooding frequencies projected for FRIs in Canada.

RCP	100-Year			250-Year		
	Increase	Decrease	No Change	Increase	Decrease	No Change
			All GCM median			
RCP 2.6	532	518	22	538	509	25
RCP 4.5	639	403	30	623	422	27
RCP 6.0	520	523	29	507	544	21
RCP 8.5	441	485	146	402	543	127
			Robust GCM median			
RCP 2.6	543	518	11	553	509	10
RCP 4.5	660	403	9	641	422	9
RCP 6.0	540	523	9	519	544	9
RCP 8.5	578	485	9	520	543	9

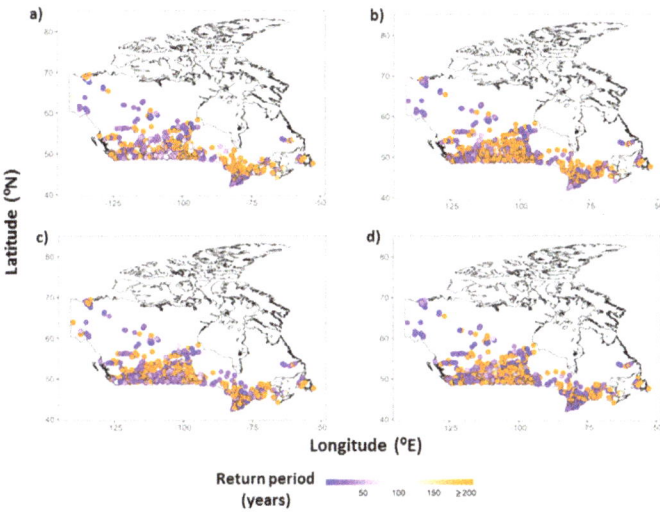

Figure 4. Updated flood frequency of historical 100-year flood magnitudes at FRIs in Canada for (**a**) RCP 2.6, (**b**) RCP 4.5, (**c**) RCP 6.0, and (**d**) RCP 8.5 obtained from robust GCM median approach.

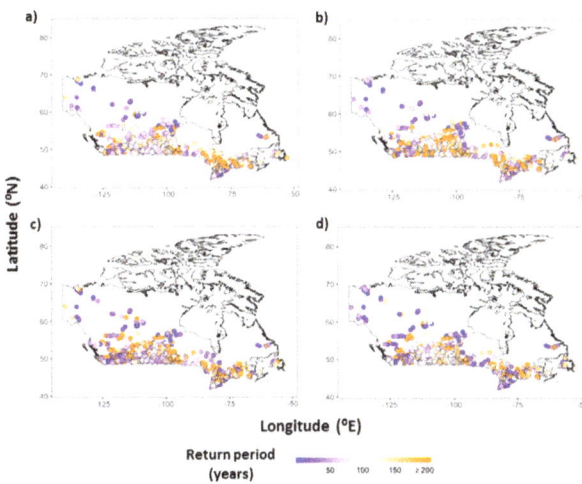

Figure 5. Future projected flooding frequencies of historical 100-year flood magnitudes at FRIs in Canada for (**a**) RCP 2.6, (**b**) RCP 4.5, (**c**) RCP 6.0, and (**d**) RCP 8.5 obtained from all GCM median.

Table 7. List of 10 FRIs projected with largest increases in flooding frequencies for historical 100-year return period flooding events from robust GCM median approach. Information on the province and projected future flood frequency magnitudes in years is also provided within brackets.

S.No.	RCP 2.6	RCP 4.5	RCP 6.0	RCP 8.5
1	Reindeer River Above Devil Rapids (SK; 2)	Playgreen Lake At Entrance To East Nelson River (MB; 2)	Reindeer River Above Devil Rapids (SK; 2)	St. Marys River At Sault Ste. Marie (Above) (ON; 2)
2	Churchill River Above Leaf Rapids (MB; 2)	Sipiwesk Lake At Sipiwesk Landing (MB; 2)	Churchill River Above Leaf Rapid (MB; 2)	St. Marys River At Sault Ste. Marie (Below) (ON; 2)
3	Churchill River Below Fidler Lake (MB; 2)	Split Lake At Split Lake (MB; 2)	Churchill River Below Fidler Lake (MB; 2)	Churchill River Above Leaf Rapid (MB; 2)
4	Mackenzie River At Arctic Red River (NT; 2)	Nelson River At Kettle Generating Station (MB; 2)	Peace River Below Chenal Des Quatre Fourches (AB; 2)	Churchill River Below Fidler Lake (MB; 2)
5	Mackenzie River At Confluence East Channel (NT; 2)	Churchill River Below Fidler Lake (MB; 2)	Riviere Des Rochers Above Slave River (AB; 2)	Peace River Below Chenal Des Quatre Fourches (AB; 2)
6	Mackenzie River At Fort Good Hope (NT; 2)	Mackenzie River At Arctic Red River (NT; 2)	Mackenzie River At Arctic Red River (NT; 2)	Lake Athabasca Near Crackingstone Point (SK; 2)
7	Playgreen Lake At Entrance To East Nelson River (MB; 2.5)	Mackenzie River At Confluence East Channel (NT; 2)	Mackenzie River At Confluence East Channel (NT; 2)	Riviere Des Rochers Above Slave River (AB; 2)
8	Sipiwesk Lake At Sipiwesk Landing (MB; 2.5)	Mackenzie River At Fort Good Hope (NT; 2)	Mackenzie River At Fort Good Hope (NT; 2)	Riviere Des Rochers East Of Little Rapids (AB; 2)
9	Mackenzie River (Peel Channel) Above Aklavik (NT; 2.5)	Mackenzie River (Peel Channel) Above Aklavik (NT; 2)	Mackenzie River (Peel Channel) Above Aklavik (NT; 2)	Riviere Des Rochers West Of Little Rapids (AB; 2)
10	Cedar Lake Near Oleson Point (MB; 3)	Churchill River Above Leaf Rapids (MB; 2.5)	St. Marys River At Sault Ste. Marie (Above) (ON; 3)	Mackenzie River At Arctic Red River (NT; 2)

Finally, projected changes in the flood timings across the FRIs are also assessed. The results are summarized in Table 9 and Figures 6 and 7. In Table 9, the total numbers of FRIs that are projected with change or no-change in flood timing are summarized along with the FRIs where uncertain flood timings are obtained for either historical or future time-periods. From the results obtained from all GCM median approach, more numbers of FRIs are projected with no-change in flood timing than FRIs projected with a change in flood timing for RCP 2.6 and RCP 4.5 whereas the opposite is noted in case of more severe emission scenarios i.e., RCP 6.0 and RCP 8.5. This suggests that a larger number of FRIs can be expected to experience changes in flood timings under more severe emission scenarios than the low and moderate ones. Figure 6 suggests that a large number of FRIs projected with changes in flood timings are located in southern Ontario, southern prairies, and along the west coast of Canada whereas FRIs projected with no change in flood timings are distributed along the east coast, northern Ontario, and northern prairies regions of Canada.

Table 8. List of 10 FRIs projected with largest increases in flooding frequencies for historical 100-year return period flooding events from all GCM median approach. Information on the province and projected future flood frequency magnitudes in years is also provided within brackets.

S.No	RCP 2.6	RCP 4.5	RCP 6.0	RCP 8.5
1	Mackenzie River At Fort Good Hope (NT; 2)	Churchill River Below Fidler Lake (MB; 2)	Reindeer River Above Devil Rapids (SK; 2)	St. Marys River At Sault Ste. Marie (Above) (ON; 2)
2	Churchill River Above Leaf Rapids (MB; 2.5)	Mackenzie River At Arctic Red River (NT; 2)	Churchill River Above Leaf Rapids (MB; 2)	St. Marys River At Sault Ste. Marie (Below) (ON; 2)
3	Churchill River Below Fidler Lake (MB; 2.5)	Mackenzie River At Confluence East Channel (NT; 2)	Peace River Below Chenal Des Quatre Fourches (AB; 2)	Churchill River Above Leaf Rapids (MB; 2)
4	Playgreen Lake At Entrance To East Nelson River (MB; 3)	Mackenzie River At Fort Good Hope (NT; 2)	Riviere Des Rochers Above Slave River (AB; 2)	Churchill River Below Fidler Lake (MB; 2)
5	Peace River Below Chenal Des Quatre Fourches (AB; 3)	Mackenzie River (Peel Channel) Above Aklavik (NT; 2)	Mackenzie River At Arctic Red River (NT; 2)	Peace River Below Chenal Des Quatre Fourches (AB; 2)
6	Riviere Des Rochers Above Slave River (AB; 3)	Reindeer River Above Devil Rapids (SK; 3)	Mackenzie River At Confluence East Channel (NT; 2)	Riviere Des Rochers Above Slave River (AB; 2)
7	Mackenzie River At Arctic Red River (NT; 3)	Churchill River Above Leaf Rapids (MB; 3)	Mackenzie River At Fort Good Hope (NT; 2)	Mackenzie River At Arctic Red River (NT; 2)
8	Mackenzie River At Confluence East Channel (NT; 3)	Peace River Below Chenal Des Quatre Fourches (AB; 3)	Mackenzie River (Peel Channel) Above Aklavik (NT; 2)	Mackenzie River At Confluence East Channel (NT; 2)
9	Mackenzie River (Peel Channel) Above Aklavik (NT; 3)	Riviere Des Rochers Above Slave River (AB; 3)	St. Marys River At Sault Ste. Marie (Above) (ON; 3)	Mackenzie River At Fort Good Hope (NT; 2)
10	Sipiwesk Lake At Sipiwesk Landing (MB; 3.5)	Lake Athabasca Near Crackingstone Point (SK; 4)	St. Marys River At Sault Ste. Marie (Below) (ON; 3)	Mackenzie River (Peel Channel) Above Aklavik (NT; 2)

Table 9. Total numbers of FRIs (out of 1072) projected with a change in flood timing in Canada.

RCP	All GCM Median			Robust GCM Median		
	Change	No Change	Uncertain	Change	No Change	Uncertain
RCP 2.6	417	655	0	126	352	594
RCP 4.5	510	562	0	124	167	781
RCP 6.0	638	434	0	209	178	685
RCP 8.5	676	396	0	115	83	874

Flood timing results obtained from robust GCM median approach and shown in Table 9 suggest that a large number of FRIs are associated with uncertain flood timings. Similar to the results from all GCM median approach, the total numbers of FRIs projected with change in flood timings are smaller than FRIs projected with no-change for RCP 2.6 and RCP 4.5 whereas they are larger for RCP 6.0 and RCP 8.5. From Figure 7 it can also be noted that FRIs located in the western prairies regions, British Columbia province, and east coast regions are associated with large uncertainties in simulated flood timing. On the other hand, FRIs located in eastern prairies regions, southern Ontario, and northern regions are reliably projected with earlier shifts in flood timing ranging 1–2 months.

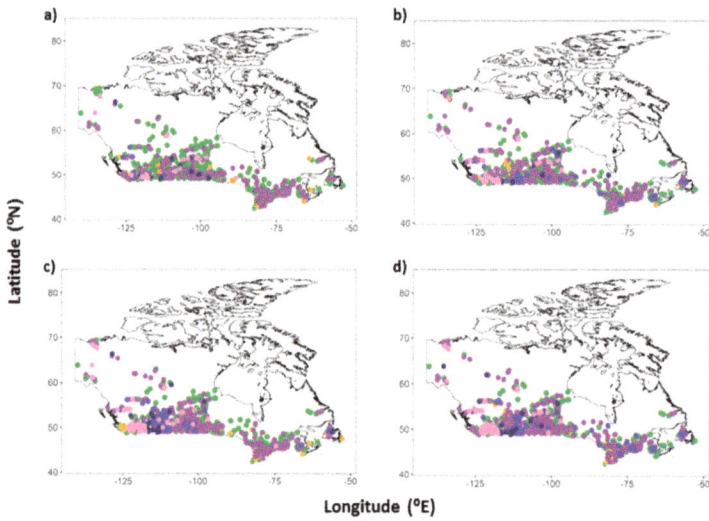

Figure 6. The change in flood timing obtained from all GCM median approach at FRIs under (**a**) RCP 2.6, (**b**) RCP 4.5, (**c**) RCP 6.0, and (**d**) RCP 8.5. FRIs projected with no change in flood timing are shown in green. FRIs projected with earlier occurrences of floods are shown in shades of blue (purple for 1 month shift, blue for 2 months shift, and dark blue for 3 month shift). FRIs projected with later occurrences of floods are shown in shades of orange (1–3 month shifts shown in light to dark orange). Finally, FRIs where 4 or more months of changes (in either direction) are projected, are shown in pink.

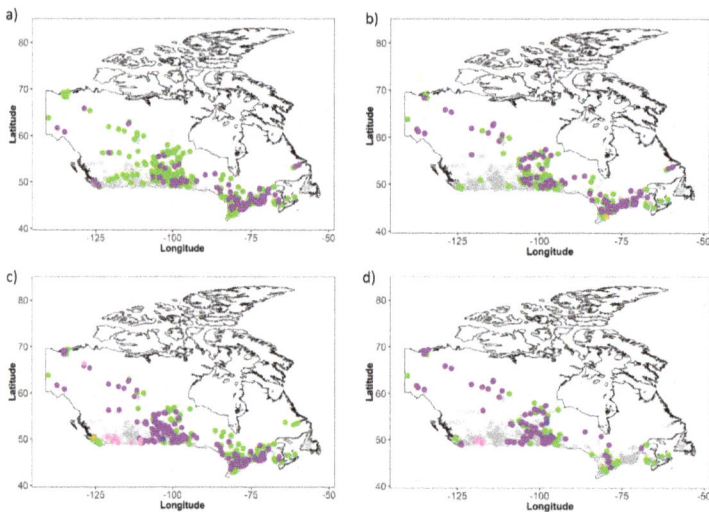

Figure 7. The change in flood timing obtained from robust GCM median approach at FRIs under (**a**) RCP 2.6, (**b**) RCP 4.5, (**c**) RCP 6.0, and (**d**) RCP 8.5. FRIs projected with no change in flood timing are shown in green. FRIs projected with earlier occurrences of floods are shown in shades of blue (purple for 1 month shift, blue for 2 months shift, and dark blue for 3 month shift). FRIs projected with later occurrences of floods are shown in shades of orange (1–3 month shifts shown in light to dark orange). FRIs where 4 or more months of changes (in either direction) are projected, are shown in pink. Finally, FRIs projected with uncertain flood timing in either historical or future timelines are shown in grey.

5. Conclusions

This study discussed the flooding related consequences of climate change on 100 most populous Canadian cities and Flow Regulation Infrastructure (FRIs) located in Canada. The aggregated results of future projected changes in flooding frequencies of historical 100 and 250 year flooding events, and changes in flood timing generated discussed in Gaur et al. [16] are used as the basis for the discussion presented in this study. Two different approaches were used to aggregate the projected changes: (1) all GCM median approach where changes projected by all GCMs (for a particular RCP) were taken into consideration and their median is calculated as the aggregated value, and (2) robust GCM median approach where changes projected by robust GCMs were considered for aggregation. More details on the procedure used to calculate the aggregated results can be obtained from Gaur et al. [16].

The projected changes from both aggregation approaches are used to assess changes in future flooding frequencies at the cities, and calculate flood hazard indices and flood risk indices to evaluate flooding related consequences of climate change on the cities. Overall, it is obtained that 40–60% of the cities are projected with increasing flooding frequencies in the future. The number of cities projected with increasing flooding frequencies, increased flood hazard index, and increased flood risk was found to be greatest under the more intense emission scenarios: RCP 6.0 and RCP 8.5 than the conservative scenarios: RCP 2.6 and RCP 4.5. In terms of chosen hazard and risk indices, it is especially noted that many cities located in southern Ontario such as Sault Ste. Marie, Nanticoke, Cornwall etc. are associated with very high flood hazard whereas other highly populated cities such as Toronto, Montreal, and Saint Catharines-Niagara are associated with the highest flood risk. Some heavily populated cities such as Ottawa and Edmonton have also been projected with future increases in flooding frequencies, decreasing flood hazard and flood risk. Large uncertainty in the sign of change as contributed by differences in RCPs is also evident in the results for some cities such as Vancouver, Quebec etc.

The flooding related consequences of climate change on FRIs are evaluated in terms of the projected changes in flooding frequencies and flood timing at their locations. Results indicate that overall 40–50% FRIs especially those located in the south-western Ontario, west coast, and northern regions of Canada can experience future increases in flooding frequencies as a consequence of climate change. The FRIs projected to experience the largest increases in flooding frequencies are found to be located in the prairies provinces such as Saskatchewan, Manitoba and Alberta, northern provinces such as Nunavut, and Ontario. In terms of flood timing, the projected results were found to differ considerably between the two aggregation approaches where the results from robust GCM median approach allocated uncertain flood timing in a larger number of FRIs. However from both the aggregation approaches, a larger number of FRIs especially those located in the southern Ontario, southern prairies, and along the western coast of Canada were found to be projected with a change in flood timing under more extreme emission scenarios: RCP 6.0 and RCP 8.5 than low and moderate emission scenarios: RCP 2.6 and RCP 4.5.

It is also worth pointing out some of the limitations and future work of this study. The results discussed in this study are based on the aggregated projections from multiple GCMs and no attempt has been made to evaluate the reliability of runoff projections made by them. However, in line with the recommendations made in previous studies (such as Knutti et al. [39]), a very large ensemble of future runoff projections have been used for analysis so that the recommendations made are robust given the limitations and uncertainty in making future projections of runoff. The estimation of flood quantiles for historical and projected future time-periods has been made considering that flooding behavior is stationary within the two time-periods however extreme value distribution methods accounting for non-stationarity exist [40] and can be used in future studies to further validate the findings of this study. Furthermore, this study only evaluates hazards and risks originating from riverine floods. Other categories of floods originating in the coastal areas due to tidal effects and sea-level rise, short duration extreme precipitation, ice-jams, and tsunamis have not been considered. Concurrency of some of these flood generating mechanisms have the potential to generate much larger flood hazard than that discussed in this study. Nevertheless, the results discussed in this paper provide novel information that will help water resource managers and policy makers to more effectively manage Canadian cities and water resource management infrastructure in Canada.

Author Contributions: A.G. (Ayushi Gaur) conceptualized the research, performed the formal analysis, and wrote the first draft of the paper. A.G. (Abhishek Gaur), D.Y., and S.P.S. provided feedback on the research approach, and reviewed and edited the first draft of the paper. All authors revised the paper and agreed on the final version of the paper.

Funding: Funding for this research came from Chaucer Syndicates (London, UK) and the Natural Sciences and Engineering Research Council of Canada (NSERC).

Acknowledgments: Authors acknowledge the financial support of Chaucer Syndicates (London, UK) and Natural Sciences and Engineering Research Council of Canada (NSERC) through the collaborative research grant awarded to the last author.

Conflicts of Interest: The authors declare no conflict of interest.

References

1. IPCC. Summary for Policymakers. In: Climate Change 2013: The Physical Science Basis. In *Contribution of Working Group I to the Fifth Assessment Report of the Intergovernmental Panel on Climate Change*; Stocker, T.F., Qin, D., Plattner, G.-K., Tignor, M., Allen, S.K., Boschung, J., Nauels, A., Xia, Y., Bex, V., Midgley, P.M., Eds.; Cambridge University Press: Cambridge, UK; New York, NY, USA, 2013.
2. IPCC. Managing the Risks of Extreme Events and Disasters to Advance Climate Change Adaptation. In *A Special Report of Working Groups I and II of the Intergovernmental Panel on Climate Change*; Field, C.B., Barros, V., Stocker, T.F., Qin, D., Dokken, D.J., Ebi, K.L., Mastrandrea, M.D., Mach, K.J., Plattner, G.K., Allen, S.K., et al., Eds.; Cambridge University Press: Cambridge, UK; New York, NY, USA, 2012; p. 582.
3. Watt, W.E.; Lathem, K.W.; Neill, C.R.; Richards, T.L.; Rousselle, J. *Hydrology of Floods in Canada: A Guide to Planning and Design*; National Research Council of Canada: Ottawa, ON, Canada, 1989.
4. Andrews, J. *Flooding: Canada Water Year Book*; Ministry of Supply and Services: Ottawa, ON, USA, 1993.
5. Burn, D.H.; Elnur, M.A. Detection of hydrological trends and variability. *J. Hydrol.* **2002**, *255*, 107–122. [CrossRef]
6. Burn, D.H.; Fan, L.; Bell, G. Identification and quantification of streamflow trends on the Canadian Prairies. *Hydrol. Sci.* **2008**, *53*, 538–549. [CrossRef]
7. Déry, S.J.; Stahl, K.; Moore, R.D.; Whitfield, P.H.; Menounos, B.; Burford, J.E. Detection of runoff timing changes in pluvial, nival and glacial rivers of western Canada. *Water Resour. Res.* **2009**, *45*, W04426. [CrossRef]
8. Burn, D.H.; Sharif, M.; Zhang, K. Detection of trends in hydrological extremes for Canadian watersheds. *Hydrol. Process.* **2010**, *24*, 1781–1790. [CrossRef]
9. Sandink, D. *Urban Flooding in Canada*; Institute for Catastrophic Loss Reduction: London, ON, Canada, 2013; Volume 52, pp. 1–94.
10. Gaur, A.; Simonovic, S.P. Projected Changes in the Dynamics of Flood Hazard in the Grand River Basin, Canada. *Br. J. Environ. Clim. Chang.* **2015**, *5*, 37–51. [CrossRef]
11. Fiorentino, M.; Gioia, A.; Iacobellis, V.; Manfreda, S. Regional analysis of runoff thresholds behaviour in Southern Italy based on theoretically derived distributions. *Adv. Geosci.* **2011**, *26*, 139–144. [CrossRef]
12. Gioia, A.; Manfreda, S.; Iacobellis, V.; Fiorentino, M. Performance of a theoretical model for the description of water balance and runoff dynamics in Southern Italy. *J. Hydrol. Eng.* **2014**, *19*, 1113–1123. [CrossRef]
13. Devkota, L.P.; Gyawali, D.R. Impacts of climate change on hydrological regime and water resources management of the Koshi River Basin, Nepal. *J. Hydrol. Reg. Stud.* **2015**, *4B*, 502–515. [CrossRef]
14. Qin, X.S.; Lu, Y. Study of Climate Change Impact on Flood Frequencies: A Combined Weather Generator and Hydrological Modeling Approach. *J. Hydrometeorol.* **2014**, *15*, 1205–1219. [CrossRef]
15. Das, S.; Millington, N.; Simonovic, S.P. Distribution choice for the assessment of design rainfall for the city of London (Ontario, Canada) under climate change. *Can. J. Civ. Eng.* **2012**, *40*, 121–129. [CrossRef]
16. Gaur, A.; Gaur, A.; Simonovic, S.P. Future Changes in Flood Hazards across Canada under a Changing Climate. *Water* **2018**, *10*, 1441. [CrossRef]

17.	Ranger, N.; Hallegatte, S.; Bhattacharya, S.; Bachu, M.; Priya, S.; Dhore, K.; Rafique, F.; Mathur, P.; Naville, N.; Henriet, F.; et al. An assessment of the potential impact of climate change on flood risk in Mumbai. *Clim. Chang.* **2011**, *104*, 139–167. [CrossRef]

18.	Kundzewicz, Z.; Kanae, S.; Seneviratne, S.; Handmer, J.; Nicholls, N.; Peduzzi, P.; Mechler, R.; Bouwer, L.; Arnell, N.; Mach, K.; et al. Flood risk and climate change: Global and regional perspectives [le risque d'inondation et les perspectives de changement climatique mondial et regional]. *Hydrol. Sci. J.* **2014**, *59*, 1–28. [CrossRef]

19.	Pappenberger, F.; Dutra, E.; Wetterhall, F.; Cloke, H.L. Deriving global flood hazard maps of fluvial floods through a physical model cascade. *Hydrol. Earth Syst. Sci.* **2012**, *16*, 4143–4156. [CrossRef]

20.	Hirabayashi, Y.; Mahendran, R.; Koirala, S.; Konoshima, L.; Yamazaki, D.; Watanabe, S.; Kanae, S. Global flood risk under climate change. *Nat. Clim. Chang.* **2013**, *3*, 816–821. [CrossRef]

21.	Winsemius, H.C.; Van Beek, L.P.H.; Jongman, B.; Ward, P.J.; Bouwman, A. A framework for global river flood risk assessments. *Hydrol. Earth Syst. Sci.* **2013**, *17*, 1871–1892. [CrossRef]

22.	Yamazaki, D.; Lee, H.; Alsdorf, E.; Dutra, E.; Kim, H.; Kanae, S.; Oki, T. Analysis of the water level dynamics simulated by a global river model: A case study in the Amazon River. *Water Resour. Res.* **2012**, *48*, W09508. [CrossRef]

23.	Dankers, R.; Feyen, L. Climate change impact on flood hazard in Europe: An assessment based on high resolution climate simulations. *J. Geophys. Res.* **2008**, *113*, D19105. [CrossRef]

24.	Yamazaki, D.; Kanae, S.; Kim, H.; Oki, T. A physically based description of floodplain inundation dynamics in a global river routing model. *Water Resour. Res.* **2011**, *47*, 1–21. [CrossRef]

25.	Jongman, B.; Ward, P.J.; Aerts, J.C.J.H. Global exposure to river and coastal flooding: Long term trends and changes. *Glob. Environ. Chang.* **2012**, *22*, 823–835. [CrossRef]

26.	Peduzzi, P.; Dao, H.; Herold, C.; Mouton, F. Assessing global exposure and vulnerability towards natural hazards: The Disaster Risk Index. *Nat. Hazards Earth Syst. Sci.* **2009**, *9*, 1149–1159. [CrossRef]

27.	de Moel, H.; Aerts, J.C.J.H.; Koomen, E. Development of flood exposure in the Netherlands during the 20th and 21st century. *Glob. Environ. Chang.* **2011**, *21*, 620–627. [CrossRef]

28.	Kleinen, T.; Petschel-Held, G. Integrated assessment of changes in flooding probabilities due to climate change. *Clim. Chang.* **2007**, *81*, 283–312. [CrossRef]

29.	Feyen, L.; Barredo, J.I.; Dankers, R. Implications of global warming and urban land use change on flooding in Europe. In *Water and Urban Development Paradigms—Towards an Integration of Engineering, Design and Management Approaches*; Feyen, J., Shannon, K., Neville, M., Eds.; CRC Press: London, UK, 2009; pp. 217–225.

30.	Feyen, L.; Dankers, R.; Bódis, K.; Salamon, P.; Barredo, J.I. Fluvial flood risk in Europe in present and future climates. *Clim. Chang.* **2012**, *112*, 47–62. [CrossRef]

31.	Van Vuuren, D.P. The representative concentration pathways: An overview. *Clim. Chang.* **2011**, *109*, 5–31. [CrossRef]

32.	Yamazaki, D.; Oki, T.; Kanae, S. Deriving a global river network map and its sub-grid topographic characteristics from a fine-resolution flow direction map. *Hydrol. Earth Syst. Sci.* **2009**, *13*, 2241–2251. [CrossRef]

33.	Ikeuchi, H.; Hirabayashi, Y.; Yamazaki, D.; Kiguchi, M.; Koirala, S.; Nagano, T.; Kotera, A.; Kanae, S. Modeling complex flow dynamics of fluvial floods exacerbated by sea level rise in the Ganges-Brahmaputra-Meghna delta. *Environ. Res. Lett.* **2015**, *10*, 124011. [CrossRef]

34.	Hu, X.; Hall, J.W.; Shi, P.; Lim, W.H. The spatial exposure of the Chinese infrastructure system to flooding and drought hazards. *Nat. Hazards* **2016**, *80*, 1083–1118. [CrossRef]

35.	Mateo, C.M.; Hanasaki, N.; Komori, D.; Tanaka, K.; Kiguchi, M.; Champathong, M.; Sukhapunnaphan, T.; Yamazaki, D.; Oki, T. Assessing the impacts of reservoir operation to floodplain inundation by combining hydrological, reservoir management, and hydrodynamic models. *Water Resour. Res.* **2014**, *50*, 7245–7266. [CrossRef]

36.	Koirala, S.; Hirabayashi, Y.; Mahendran, R.; Kanae, S. Global assessment of agreement among streamflow projections using CMIP5 model outputs. *Environ. Res. Lett.* **2014**, *9*, 064017. [CrossRef]

37.	Takata, K.; Emori, S.; Watanabe, T. Development of the minimal advanced treatments of surface interaction and runoff. *Glob. Planet. Chang.* **2003**, *38*, 209–222. [CrossRef]

38.	Kim, H.; Yeh, P.; Oki, T.; Kanae, S. Role of rivers in the seasonal variations of terrestrial water storage over global basins. *Geophys. Res. Lett.* **2009**, *36*, L17402. [CrossRef]

39. Knutti, R.; Furrer, R.; Tebaldi, C.; Cermak, J.; Meehl, G.A. Challenges in Combining Projections from Multiple Climate Models. *J. Clim.* **2010**, *23*, 2739–2758. [CrossRef]
40. Salas, J.D.; Obeysekera. Revisiting the Concepts of Return Period and Risk for Nonstationary Hydrologic Extreme Events. *J. Hydrol. Eng.* **2014**, *19*, 554–568. [CrossRef]

water

MDPI

Communication

The Future of Drought in the Southeastern U.S.: Projections from Downscaled CMIP5 Models

David Keellings * and Johanna Engström

Department of Geography, University of Alabama, 204 Farrah Hall, P.O. Box 870322, Tuscaloosa, AL 35487-0322, USA; jengstrom@ua.edu
* Correspondence: djkeellings@ua.edu; Tel.: +1-205-348-4942

Received: 27 December 2018; Accepted: 29 January 2019; Published: 2 February 2019

Abstract: After being repeatedly struck by droughts in the last few decades, water managers and stakeholders in the Southeast U.S. dread the future extremes that climate change might cause. In this study, the length of future dry periods is assessed using a sub-ensemble of downscaled CMIP5 climate models, which are proven to perform well in precipitation estimations. The length of a dry spell with a twenty-year return period is estimated for the cold and warm seasons for two time periods; 2020–2059 and 2060–2099, and considering two emission scenarios: RCP 4.5 and 8.5. The estimates are then compared with historical dry spells and differences in length and geospatial distribution analyzed. Based on the findings of this paper, little change can be expected in dry spell length during the warm season. Greater changes are to be expected in the cold season in the southern half of Florida, where dry spells are expected to be up to twenty days shorter, while dry spells in Alabama, Mississippi and Tennessee are predicted to be up to twenty days longer. The changes predicted by the models are positively associated with emission trajectory and future time period.

Keywords: consecutive dry days; climate; downscaled projections; Southeast U.S.; CMIP5

---

1. Introduction

The Southeast United States (hereafter referred to as "the Southeast") is a region traditionally subject to abundant precipitation and water resources, brought by local convection and tropical cyclones during the warm season, and cold fronts during the cool season [1]. In recent decades, the region has been subject to multiple droughts (1986–1988, 1998–2002, 2007–2008, 2016). These droughts typically don't last as long as droughts experienced in the Southwest United States, but they have proved to be highly detrimental to the regional environment, economy, and political situation [2–4]. Some of the region's hydroclimatological variability has been attributed to large-scale atmospheric drivers or teleconnections, such as El Niño-Southern Oscillation, the Atlantic Multidecadal Oscillation and the North Atlantic Oscillation [5–10] and more recently, baroclinic instabilities over the oceans [11]. However, due to the severity and frequency of recent droughts there is a need among water managers and policy makers to look beyond the seasonal forecasts, afforded by teleconnections, to what the future holds for the Southeast over the longer term.

The IPCC [12] predicts that the annual precipitation amounts for the Southeast might increase by up to 20% by year 2035, while evaporation rates are expected to increase by up to 5%, but does not specifically mention dry spells. However, many models do project drier conditions in the far southwest of the Southeast region and wetter conditions in the far northeast of the region [13]. The U.S. National Climate Assessment finds with a high degree of certainty that both extreme wetness and dryness are projected to increase across the U.S. by the end of this century. The report finds that annual maximum precipitation will increase by up to 30% across much of the U.S. and length of consecutive dry days will increase by up to 30%, particularly in the southern U.S. [1]. Estimates of the length of the longest annual consecutive dry period are expected to increase by up to five days in the Southeast, while there are no significant trends observed in the number of heavy rain days [14].

Previous work has found increased future lengths of Consecutive Dry Days (CDD) in many regions of the world based on projections from multi-model CMIP5 ensembles [14,15]. However, relatively little change in consecutive dry days has been projected for the Southeast U.S. [15]. These studies have also suggested that increases in consecutive dry days coincide with increased precipitation intensity. For example, small decreases in future wet day probability have been found in the Southeast during December, January, February and June, July, August with the largest reductions occurring in coastal areas and peninsular Florida [15]. These findings were accompanied by projections of increased daily intensity of precipitation. These projected changes suggest longer periods of dry weather punctuated by more intense precipitation events. The findings are consistent with those of another study [16] that reported upward trends in the frequency and intensity of observed intense precipitation events across the Southeast associated with more frequent northward intrusion of moist tropical air masses, particularly in summer months.

Analyses of paleodata has shown that in the Southeast the last century has been among the wettest periods in the historic record, dating back a few centuries [1,17,18]. Yet regional water managers have expressed concerns about what's perceived as drier-than average conditions prevailing during the last few decades and they worry about what future climate conditions will bring. Although there is variability across future projections of hydroclimate in the Southeast, the projections are generally moderate compared to many midlatitude regions around the globe that are expected to face more extreme climatic conditions as a result of climate change. Hence, studies investigating the impact of climate change in the Southeast have been limited.

Here, seasonal projections of 20-year return period lengths of consecutive dry days are examined from an ensemble of downscaled global climate models and compared with historical observations. The ensemble of downscaled models is comprised of those selected from a suite of 32 models, based on performance in simulation of observed consecutive dry spells, identified in an earlier study by the authors [19]. This is essentially a weighted ensemble approach [20–23] which removes the influence of under-performing models. Adding to the novelty of the methods presented here is the use of the relatively new downscaling technique of localized constructed analogs (LOCA) [24].

2. Materials and Methods

2.1. Data

Thirty-two LOCA downscaled ($1/16°$) spatial resolution Global Climate Models from the World Climate Research Program's (WCRP) Coupled Model Intercomparison Project phase 5 (CMIP5) were downloaded from the Downscaled CMIP3 and CMIP5 Climate and Hydrology Projections archive, (http://gdo-dcp.ucllnl.org/downscaled_cmip_projections/dcpInterface.html). Future LOCA downscaled runs of these models are subdivided into the near future (2020–2059) and future (2060–2099). LOCA is distinct from other constructed analog techniques in that it finds a single best match analog day for each point instead of using weighted sums and thus gives a better estimation of extremes [24]. Each of the 32 models has one LOCA ensemble member. Two representative concentration pathways (RCPs) are used to represent different greenhouse gas (GHG) emissions trajectories: rapid GHG emission growth RCP 8.5 (8.5 W/m^2 radiative forcing, ~1370 ppm CO$_2$), medium GHG emission growth with stabilization RCP 4.5 (4.5 W/m^2 radiative forcing, ~650 ppm CO$_2$) [25].

Observed precipitation data is a gridded dataset which is a subset of an interpolation of observed precipitation data from 20,000 National Oceanic and Atmospheric Administration (NOAA) Cooperative Observer Network (COOP) stations, and part of the North American Land Data Assimilation System Variable Infiltration Capacity simulations over North America (http://www.colorado.edu/lab/livneh/data) [26–29]. These daily data (1950–2005) have the same temporal and spatial resolution as the LOCA data and are used for historical comparison. Climatic regions within the United States can be defined in numerous ways that vary with the scope of the analysis [1,8,30,31]. In this paper the Southeast is defined as the southern states east of the Mississippi river and includes Alabama, Florida, Georgia, Mississippi, North Carolina, South Carolina and Tennessee (Figure 1). The results are presented by climate division (nClimDiv, provided by NOAA) within the Southeast. A dry day is defined according to the definition of no measurable precipitation, as a day with less than 3 mm of precipitation [1,19,32].

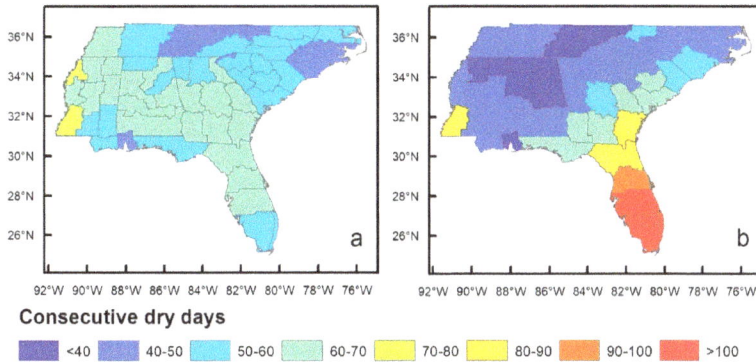

Figure 1. Observed (1950–2005) lengths of 20-year return period consecutive dry days. For warm season (**a**) and cold season (**b**).

These data are divided into two seasons based on agricultural growing dates. The warm season is defined as April–October, while the cold season constitutes the months of November–March. These seasons are based on the U.S. Department of Agriculture's data on planting and harvesting dates of field crops in the region [33]. A severe dry spell during the warm season could be detrimental for agriculture and the environment, while a cool season drought limits the possibilities for recharge and storage of water in both natural and anthropogenic reservoirs.

2.2. Model Evaluation

The performance of all 32 LOCA downscaled CMIP5 models in simulation of observed counts of CDD were evaluated using probability density function comparisons by way of the Perkins skill scores [34] of the distribution of dry days as well as the upper tail (extreme-based) of the distribution. The Keellings significance test was used to assess the significance of the Perkins skill scores [30]. These methods, as applied to assessment of the 32 LOCA downscaled model simulations of CDD in the Southeast and resulting model performance are fully described in previous work [19]. Here we only examine future runs of the top five best performing models as identified in each season (Table 1) [19]. The skill scores of all 32 models averaged across all climate divisions ranged between 0.91 and 0.93. Model performance was heterogeneous across the region, but models generally performed better in the cold season than in the warm season.

The Generalized Extreme Value (GEV) distribution is used to estimate return periods of lengths of dry spells between observations and models [35–37]. The GEV is utilized as it makes no a priori assumptions regarding the form of the extreme value distribution [38]. Here we estimate the GEV parameters with the method of maximum likelihood using the extRemes package in R. The cumulative distribution function of the GEV is given by:

$$P(x) = exp\left[-\{1 + \xi \frac{x - \mu}{\sigma}\}^{-1/\xi}\right] \qquad (1)$$

where x is the value of a random variable i.e., number of CDD, μ is the location parameter (central tendency), σ is the scale parameter (variance), and ξ is the shape parameter (skew) [39]. Future return periods are estimated using the fitted GEV for the ensemble of the best performing models, as identified by the skill score measures [19], in each future period (2020–2059, 2060–2099) and for each RCP (RCP 4.5, RCP 8.5). Future CDD return period estimates are then compared to the observation return periods.

Table 1. List of CMIP5 models evaluated in each season. Top five models of 32 as identified by application of skill scores (averaged across all climate divisions) [19].

Season	Model Name	Affiliation	Country	Skill
Warm Season	ACCESS1-3	CSIRO (Commonwealth Scientific and Industrial Research Organization, Australia), and BOM (Bureau of Meteorology, Australia)	Australia	0.92
	GISS-E2-H GISS-E2-R	NASA Goddard Institute for Space Studies	United States	0.93
	MIROC5	Atmosphere and Ocean Research Institute (The University of Tokyo), National Institute for Environmental Studies, and Japan Agency for Marine-Earth Science and Technology	Japan	0.93
	MRI-CGCM3	Meteorological Research Institute	Japan	0.92
Cold Season	CanESM2	Canadian Centre for Climate Modeling and Analysis	Canada	0.93
	CESM1-CAM5	National Science Foundation, Department of Energy, National Center for Atmospheric Research	United States	0.92
	inmcm4	Institute for Numerical Mathematics	Russia	0.92
	IPSL-CM5A-LR IPSL-CM5A-MR	Institute Pierre-Simon Laplace	France	0.92

3. Results

Figure 1 shows the length of observed (1950–2005) 20-year recurring CDD for the warm season (Figure 1a) and cold season (Figure 1b). Warm season precipitation in the Southeast is generally frequent and originates from local convection as well as tropical cyclones. The observational record shows that the 20-year CDD for the warm season is up to 70 days. The cold season (Figure 1b) shows a North–South gradient in the length of dry spells. Precipitation during this time of the year mainly originates from cold fronts moving south over the continent, which translates to 20-year CDD of less than 40 days in the northern parts of the study area during the cold season, while farther south 100 plus CDD are observed.

Figure 2 shows the length of a 20-year return CDD, as estimated by the best performing CMIP5 models [19]. Visually comparing with Figure 1 (observations), Figure 2 shows generally similar spatial patterns of lengths of CDD 20-year return periods indicating that the models capture the spatial dynamics involved in physical processes responsible for CDD.

Figure 3 shows the absolute difference between the observed (1950–2005) length of a 20-year CDD and the future outlooks. Figure 3 illustrates that summertime dry spells (left column of maps) will exhibit little change throughout the region with few of the climate divisions exhibiting a change in excess of ±5 days in length of CDD. However, there is a reduction in 20-year return values by 5–10 days in approximately 25% of the climate divisions during the near future period under the high emissions scenario. The winter season (right column of Figure 3) shows greater change and a more distinct geographical pattern. There is an increase in the length of 20-year return CDD by up to 20 days in more northern and continental locations and reductions of 20 days or more in more coastal locations, especially peninsular Florida. These changes are apparent across both future periods and both emissions scenarios, but they are more pronounced in the later period and under the higher RCP 8.5.

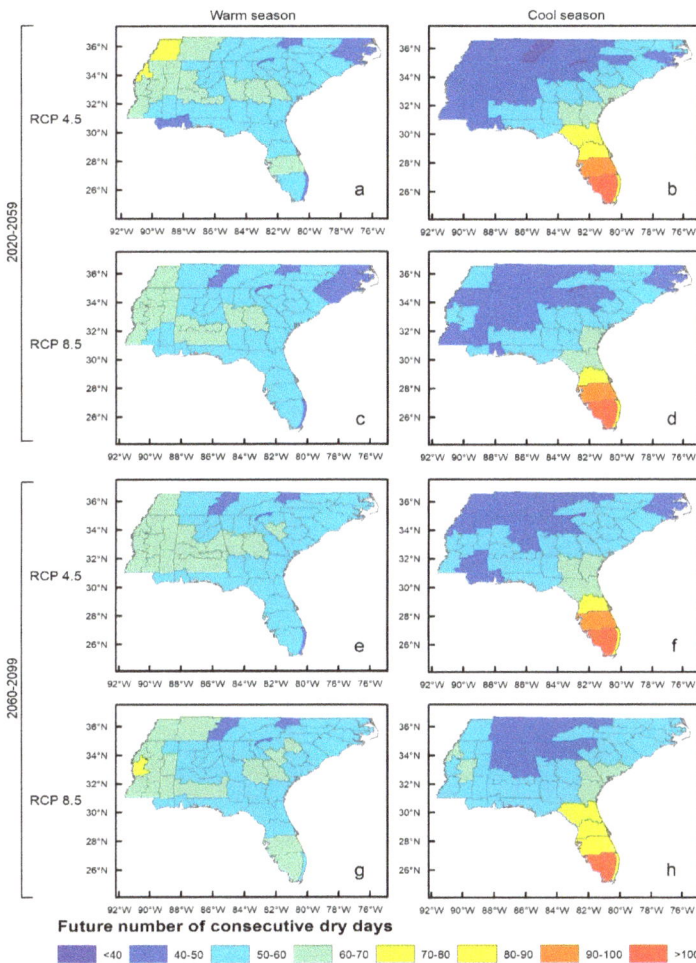

Figure 2. Modeled future length of 20-year return period consecutive dry days. Warm season estimates are shown in (**a,c,e,g**), and cold season estimates are shown in (**b,d,f,h**). Future period and RCP scenario are indicated on the left.

The southwestern-most climate division of Mississippi shows a spatially homogeneous behavior in Figure 2, but one climate division with significantly different future outlooks than surrounding climate divisions appears in Figure 3. The explanation for this is an extreme drought that struck southern Mississippi in the 1950s [40,41]. This drought can also be seen in Figure 1 where the same climate division stands out as having a considerably longer 20-year CDD than surrounding climate divisions. This results in a large difference between the length of the historical and modelled future 20-year CDD. It should also be noted that when calculating the length of 20-year historical CDD in each climate division, the maximum observed values were used. This value might represent an outlier within the climate division. As the focus of this paper is on extreme CDD, this was viewed as a preferential method to averaging the CDD values within each climate division.

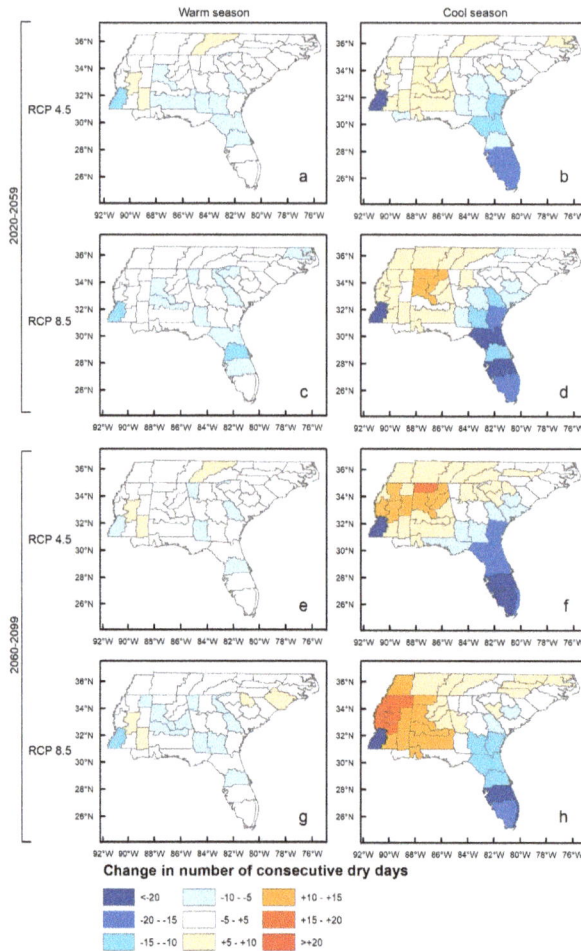

Figure 3. Absolute changes in length of 20-year return period consecutive dry days in modeled future periods versus observations. Warm season estimates are shown in (**a,c,e,g**), and cold season estimates are shown in (**b,d,f,h**). Future period and RCP scenario are indicated on the left.

4. Discussion and Conclusions

As a continuation of the authors' earlier work [19], this paper analyses meteorological drought from the perspective of consecutive dry days (days receiving less than 3 mm precipitation). The authors recognize that the nature of a drought can be multifaceted, and there can be multiple factors involved in the development of a drought, such as water use by humans, and high temperatures, leading to high water losses through evaporation and reductions in soil moisture. As has been pointed out, while consecutive dry days represent the lower tail of the precipitation distribution and are indicative of meteorological drought they should be interpreted in combination with other precipitation or drought indicators [14,42]. Thus, further work is required on how other drought indicators may change in the Southeast and how these indicators may combine with the projected changes in CDD, presented here, to increase future drought severity.

Our previous work, and that of others, has found that the CMIP5 models analyzed here can replicate the spatial patterns and magnitudes of consecutive dry days well [14,19]. However, precipitation is generally not well represented in models, particularly extremes of precipitation in the southern U.S., likely as a result of the varied physical processes responsible for precipitation formation in the region and the underestimation of tropical cyclone numbers [43]. The discrepancy between the generally poor estimation of extreme precipitation versus good estimation of CDD in the southern U.S. is likely explained by the scale of processes involved. CDD are most likely caused by larger spatial scale processes than extreme precipitation events, which are more likely the consequence of localized or mixed processes and thus more difficult for models to resolve [14]. Further work is needed to assess the possible drivers of the results presented here which are likely a combination of changes in atmospheric moisture and circulation.

In this paper, we have presented projections of extreme CDD lengths across the Southeast using a small ensemble of high performing models. These models have been downscaled to a fine spatial resolution using a technique that has been shown to preserve rather than dilute extreme values. The projections are divided into warm and cold seasons based on agricultural growing dates as to isolate seasons subject to differing societal concerns and changes in CDD are presented for two future periods, in order to differentiate between near term versus long term impacts, and for two emissions trajectories to aid in decision making. Projected CDD exhibit little change during the warm season, but during the cold season there is an increase of up to 20 days in length of 20-year return CDD in more northern and continental locations and reductions of 20 days or more in coastal locations.

The projection of greatest increase of CDD in more northern and continental locations is concerning as these areas of the Southeast coincide with locations of the largest reservoirs in the region. These surface reservoirs are regionally significant sources of drinking water and irrigation and are essential for hydropower production, which is the primary form of renewable electricity generation in the Southeast. The seasonality of the projected increase in CDD is also of concern as wildfires are generally more likely to occur during the cold season. Drought risk is likely to increase in the future throughout the region given projections of increased temperatures and rates of evapotranspiration as a result of climate change [1,44]. It is, therefore, likely that the projected increases in CDD will place further water resource stress on the Southeast.

Author Contributions: Conceptualization, J.E. and D.K., methodology, D.K.; validation, J.E. and D.K., visualization J.E. Both authors contributed equally to the writing and editing of the manuscript.

Funding: This research received no external funding.

Acknowledgments: The authors gratefully acknowledge use of the resources of the Alabama Water Institute at The University of Alabama. Further the authors acknowledge the WCRP's Working Group on Coupled Modelling, which is responsible for CMIP, and thanks to the climate modelling groups for producing and making available their model output. For CMIP, the US Department of Energy's PCMDI provides coordinating support and led development of software infrastructure in partnership with Global Organization for Earth System Science Portals. The authors also greatly appreciate the work of DCHP in their downscaling effort and for making available the LOCA downscaled model output and Ben Livneh's hydrometeorology data set.

Conflicts of Interest: The authors declare no conflict of interest.

References

1. Kunkel, K.E.; Stevens, L.E.; Stevens, S.E.; Sun, L.; Anssen, E.; Wuebbles, D.; Redmond, K.T.; Dobson, J.G. Regional Climate Trends and Scenarios for the U.S. National Climate Assessment. *NOAA Tech. Rep. NESDIS* **2013**, *142–146*, 103.
2. Yuhas, E.; Daniels, T. The US freshwater supply shortage: Experiences with desalination as part of the solution. *J. Environ. Plan. Manag.* **2006**, *49*, 571–585. [CrossRef]
3. Manuel, J. Drought in the Southeast: Lessons for Water Management. *Environ. Health Perspect.* **2008**, *116*, A168–A171. [CrossRef] [PubMed]
4. Price, K. Thirsty City: Politics, Greed, and the Making of Atlanta's Water Crisis by Skye Borden. *Southeast. Geogr.* **2017**, *57*, 322–326. [CrossRef]
5. Ropelewski, C.F.; Halpert, M.S. Global and Regional Scale Precipitation Patterns Associated with the El Nino Southern Oscillation. *Mon. Weather Rev.* **1987**, *115*, 1606–1626. [CrossRef]

6.   Dracup, J.A.; Kahya, E. The relationships between U.S. streamflow and La Nina Events. *Water Resour. Res.* **1994**, *30*, 2133–2141. [CrossRef]

7.   Enfield, D.B.; Mestas-Nunez, A.M.; Trimble, P.J. The Atlantic Multidecadal Oscillation and its relation to rainfall and river flows in the continental U.S. *Geophys. Res. Lett.* **2001**, *28*, 2077–2080. [CrossRef]

8.   Ortegren, J.T.; Knapp, P.A.; Maxwell, J.T.; Tyminski, W.P.; Soulé, P.T. Ocean-Atmosphere Influences on Low-Frequency Warm-Season Drought Variability in the Gulf Coast and Southeastern United States. *J. Appl. Meteorol. Climatol.* **2011**, *50*, 1177–1186. [CrossRef]

9.   Labosier, C.; Quiring, S. Hydroclimatology of the Southeastern USA. *Clim. Res.* **2013**, *57*, 157–171. [CrossRef]

10.   Engström, J.; Waylen, P. Drivers of long-term precipitation and runoff variability in the southeastern USA. *Theor. Appl. Climatol.* **2018**, *131*, 1133–1146. [CrossRef]

11.   Pinault, J.-L. Regions Subject to Rainfall Oscillation in the 5–10 Year Band. *Climate* **2018**, *6*, 2. [CrossRef]

12.   IPCC. *IPCC Climate Change 2014: Synthesis Report*; IPCC: Geneva, Switzerland, 2014.

13.   Carter, L.M.; Jones, J.W.; Berry, L.; Burkett, V.; Murley, J.F.; Obeysekera, J.; Schramm, P.J.; Wear, D. Chapter 17: Southeast and the Caribbean. In *Climate Change Impacts in the United States: The Third National Climate Assessment*; U.S. Global Change Research Program; U.S. Government Printing Office: Washington, DC, USA, 2014; pp. 396–417.

14.   Sillmann, J.; Kharin, V.V.; Zwiers, F.W.; Zhang, X.; Bronaugh, D. Climate extremes indices in the CMIP5 multimodel ensemble: Part 2. Future climate projections. *J. Geophys. Res. Atmos.* **2013**, *118*, 2473–2493. [CrossRef]

15.   Schoof, J.T. High-resolution projections of 21st century daily precipitation for the contiguous U.S. *J. Geophys. Res.* **2015**, *120*, 3029–3042. [CrossRef]

16.   Skeeter, W.J.; Senkbeil, J.C.; Keellings, D.J. Spatial and temporal changes in the frequency and magnitude of intense precipitation events in the southeastern United States. *Int. J. Climatol.* **2018**. [CrossRef]

17.   Seager, R.; Tzanova, A.; Nakamura, J. Drought in the Southeastern United States: Causes, Variability over the Last Millennium, and the Potential for Future Hydroclimate Change. *J. Clim.* **2009**, *22*, 5021–5045. [CrossRef]

18.   Pederson, N.; Bell, A.R.; Knight, T.A.; Leland, C.; Malcomb, N.; Anchukaitis, K.J.; Tackett, K.; Scheff, J.; Brice, A.; Catron, B.; et al. A long-term perspective on a modern drought in the American Southeast. *Environ. Res. Lett.* **2012**, *7*, 14034. [CrossRef]

19.   Engström, J.; Keellings, D. Drought in the Southeastern USA: An assessment of downscaled CMIP5 models. *Clim. Res.* **2018**, *74*, 251–262. [CrossRef]

20.   Knutti, R. The end of model democracy? *Clim. Chang.* **2010**, *102*, 395–404. [CrossRef]

21.   Räisänen, J.; Palmer, T.N. A probability and decision-model analysis of a multimodel ensemble of climate change simulations. *J. Clim.* **2001**, *14*, 3212–3226. [CrossRef]

22.   Sánchez, E.; Romera, R.; Gaertner, M.A.; Gallardo, C.; Castro, M. A weighting proposal for an ensemble of regional climate models over Europe driven by 1961–2000 ERA40 based on monthly precipitation probability density functions. *Atmos. Sci. Lett.* **2009**, *10*, 241–248. [CrossRef]

23.   Sun, F.; Roderick, M.L.; Lim, W.H.; Farquhar, G.D. Hydroclimatic projections for the Murray-Darling Basin based on an ensemble derived from Intergovernmental Panel on Climate Change AR4 climate models. *Water Resour. Res.* **2011**, *47*. [CrossRef]

24.   Pierce, D.W.; Cayan, D.R.; Thrasher, B.L. Statistical Downscaling Using Localized Constructed Analogs (LOCA). *J. Hydrometeorol.* **2014**, *15*, 2558–2585. [CrossRef]

25.   Van Vuuren, D.P.; Edmonds, J.; Kainuma, M.; Riahi, K.; Thomson, A.; Hibbard, K.; Hurtt, G.C.; Kram, T.; Krey, V.; Lamarque, J.F.; et al. The representative concentration pathways: An overview. *Clim. Chang.* **2011**, *109*, 5–31. [CrossRef]

26.   Shepard, D.S. Computer Mapping: The SYMAP Interpolation Algorithm. In *Spatial Statistics and Models*; Springer: Dordrecht, The Netherlands, 1984; ISBN 978-90-481-8385-2.

27.   Widmann, M.; Bretherton, C.S. Validation of Mesoscale Precipitation in the NCEP Reanalysis Using a New Gridcell Dataset for the Northwestern United States. *J. Clim.* **2000**, *13*, 1936–1950. [CrossRef]

28.   Maurer, E.P.; Wood, A.W.; Adam, J.C.; Lettenmaier, D.P.; Nijssen, B.; Maurer, E.P.; Wood, A.W.; Adam, J.C.; Lettenmaier, D.P.; Nijssen, B. A Long-Term Hydrologically Based Dataset of Land Surface Fluxes and States for the Conterminous United States. *J. Clim.* **2002**, *15*, 3237–3251. [CrossRef]

29. Livneh, B.; Bohn, T.J.; Pierce, D.W.; Munoz-Arriola, F.; Nijssen, B.; Vose, R.; Cayan, D.R.; Brekke, L. A spatially comprehensive, hydrometeorological data set for Mexico, the U.S., and Southern Canada 1950–2013. *Sci. Data* **2015**, *2*, 150042. [CrossRef]

30. Keellings, D. Evaluation of downscaled CMIP5 model skill in simulating daily maximum temperature over the southeastern United States. *Int. J. Climatol.* **2016**, *36*, 4172–4180. [CrossRef]

31. Engström, J.; Waylen, P. The changing hydroclimatology of Southeastern U.S. *J. Hydrol.* **2017**, *548*, 16–23. [CrossRef]

32. Wilks, D. *Statistical Methods in the Atmospheric Sciences*; Elsevier Inc.: Amsterdam, The Netherlands, 2011; ISBN 9781849733403.

33. US Department of Agriculture National Agricultural Statistics Service. Field Crops Usual Planting and Harvesting Dates. 2010. Available online: http://usda.mannlib.cornell.edu/usda/current/planting/planting-10-29-2010.pdf (accessed on 8 February 2017).

34. Perkins, S.E.; Pitman, A.J.; Sisson, S.A. Systematic differences in future 20 year temperature extremes in AR4 model projections over Australia as a function of model skill. *Int. J. Climatol.* **2013**, *33*, 1153–1167. [CrossRef]

35. Zwiers, F.W.; Kharin, V.V. Changes in the extremes of the climate simulated by CGC GCM2 under CO2 doubling. *J. Clim.* **1998**, *11*, 2200–2222. [CrossRef]

36. Kharin, V.V.; Zwiers, F.W.; Zhang, X. Intercomparison of Near-Surface Temperature and Precipitation Extremes in AMIP-2 Simulations, Reanalyses, and Observations. *J. Clim.* **2005**, *18*, 5201–5223. [CrossRef]

37. Waylen, P.; Keellings, D.; Qiu, Y. Climate and health in Florida: Changes in risks of annual maximum temperatures in the second half of the twentieth century. *Appl. Geogr.* **2012**, *33*, 73–81. [CrossRef]

38. Jenkinson, A. The frequency distribution of annual maximum (and minimum) values of meteorological elements. *Q. J. R. Meteorol. Soc.* **1955**, *87*, 158–171. [CrossRef]

39. Coles, S. *An Introduction to Statistical Modeling of Extreme Values*; Springer: London, UK, 2001; ISBN 1-85233-459-2.

40. Harvey, E.J. Records of wells in the alluvium in northwestern Mississippi. *Miss. Board Water Comm. Bull.* **1956**, *56*, 130.

41. Paulson, R. *National Water Summary 1988–1989: Hydrologic Events and Floods and Droughts*; United States Geological Survey Water-Supply Paper: Reston, VA, USA, 1991.

42. Orlowsky, B.; Seneviratne, S.I. Global changes in extreme events: Regional and seasonal dimension. *Clim. Chang.* **2012**, *110*, 669–696. [CrossRef]

43. Sheffield, J.; Barrett, A.P.; Colle, B.; Fernando, D.N.; Fu, R.; Geil, K.L.; Hu, Q.; Kinter, J.; Kumar, S.; Langenbrunner, B.; et al. North American Climate in CMIP5 experiments. Part I: Evaluation of historical simulations of continental and regional climatology. *J. Clim.* **2013**, *26*, 9209–9245. [CrossRef]

44. Sobolowski, S.; Pavelsky, T. Evaluation of present and future North American Regional Climate Change Assessment Program (NARCCAP) regional climate simulations over the southeast United States. *J. Geophys. Res. Atmos.* **2012**, *117*, D01101. [CrossRef]

water

MDPI

Article

Climate Change Impacts on Drought-Flood Abrupt Alternation and Water Quality in the Hetao Area, China

Yuheng Yang [1], Baisha Weng [1,*], Wuxia Bi [1,2], Ting Xu [1], Dengming Yan [1,3] and Jun Ma [1,4]

[1] State Key Laboratory of Simulation and Regulation of Water Cycle in River Basin, China Institute of Water Resources and Hydropower Research, Beijing 100038, China; 1109080115@cau.edu.cn (Y.Y.); biwuxia_1992@163.com (W.B.); xuting900515@163.com (T.X.); 201513408@mail.sdu.edu.cn (D.Y.); 201613332@mail.sdu.edu.cn (J.M.)

[2] College of Hydrology and Water Resources, Hohai University, Nanjing 210098, China

[3] College of Environmental Science and Engineering, Donghua University, Shanghai 201620, China

[4] School of Water Conservancy and Hydroelectric Power, Hebei University of Engineering, Handan 056021, China

* Correspondence: baishaweng@163.com; Tel.: +86-136-0598-2766

Received: 26 February 2019; Accepted: 21 March 2019; Published: 29 March 2019

Abstract: Drought-flood abrupt alternation (DFAA) is an extreme hydrological phenomenon caused by meteorological anomalies. To combat the climate change, the watershed integrated management model—Soil and Water Assessment Tool model (SWAT)—was used to simulate DFAA, total nitrogen (TN) and total phosphorus (TP) from 1961 to 2050, based on measured precipitation data in the Hetao area and the downscaled Representative Concentration Pathways (RCPs) climate scenarios. In the future, the increase in temperature and the increase in extreme precipitation will aggravate the pollution of water bodies. Results indicate that the risk of water quality exceeding the standard will increase when DFAA happens, and the risk of water quality exceeding the standard was the greatest in the case of drought-to-flood events. Results also indicate that, against the backdrop of increasing temperature and increasing precipitation in the future, the frequency of long-cycle and short-cycle drought-flood abrupt alternation index (LDFAI, SDFAI) in the Hetao area will continue to decrease, and the number of DFAA situations will decrease. However, the zone of high-frequency DFAA situations will move westward from the eastern Ulansuhai Nur Lake, continuing to pose a risk of water quality deterioration in that region. These results could provide a basis for flood control, drought resistance and pollution control in the Hetao and other areas.

Keywords: drought-flood abrupt alternation; temporal and spatial evolution; climate change; water quality; Copula function

1. Introduction

In 2018, the Intergovernmental Panel on Climate Change (IPCC) pointed out that the global average temperature has risen by nearly 1.5 °C, and it is predicted that it will increase by another 1.1–6.4 °C in 2100 [1,2]. The Fifth Assessment Report (AR5) of the IPCC stated that climate system warming is an undoubted fact [3]. Climate warming has not only directly affects extreme temperature fluctuations but also increases the frequency and intensity of extreme weather, such as, high temperature, droughts, rainstorms and floods, especially in areas sensitive and vulnerable to climate change [4–6].

DFAA is a type of extreme weather affected by climate change [7]. It refers to drought in a certain period of time and flooding in another period, an alternating occurrence of droughts and floods [8]. DFAA research increased in the 20th century. During this period, there were more extensive and thorough researches on extreme droughts and floods, of their causes, their alternating occurrence and their co-occurrence. Vogel et al. [9] studied the weather sources of abnormal precipitation in St. Louis by the METROMEX network method. Trenberth et al. [10] studied the extreme drought in 1988 and the physical causes of the extreme flood in 1993 in the United States.

The relationship between atmospheric circulation anomalies and DFAA is also the focus of scholars [11]. Garnett et al. [12] conducted a statistical analysis of the effects of El Niño-Southern Oscillation (ENSO) and Indian monsoon droughts and floods on food production using global food production data. Hastenrath et al. [13] studied atmospheric circulation mechanism anomalies in the droughts and floods change in eastern Equatorial Africa from 2005 to 2008. Chinese researchers are concentrating on the middle and lower reaches of the Yangtze River where DFAA frequently occurs. Yang et al. [7] believed that the significant differences in water vapor transport flux and the atmospheric circulation field before and after DFAA are the main reasons for DFAA in the middle and lower reaches of the Yangtze River in 2011. Zhang et al. [14] revealed that in addition to climatic factors, land-water storage anomalies and the contradiction between the characteristics of easily occurring flood (drought) and insufficient flood drainage (drought control) ability are also major causes.

To analyze the law of DFAA, historical data was applied to study its occurrence, and evaluation indicators were used to predict the future DFAA characteristics [15,16]. In terms of the evaluation indicators, the precipitation-based summer LDFAI has been widely used in predicting the turn from drought to flood as well as the turn from flood-to-drought during flood season [17,18]. Wu et al. [19,20] analyzed the characteristics of DFAA in the summer of a normal monsoon year and predicted that the total precipitation during the co-occurrence of drought and flood, and DFAA in the summer in southern China tended to be normal. Mosavi et al. [21,22] used machine learning models and hybrid neuro-fuzzy algorithms to predict the likelihood of future floods and droughts. After predicting future drought and flooding results, the sensitivity of the results should be tested. Choubin, et al. [23] used multivariate discriminant analysis, classification and regression trees, and support vector machines to perform sensitivity tests on flood prediction results. In addition, the occurrence of drought and flood in the regions affected by global climate change and human activities has been increasingly frequent, and the possibility of DFAA continues to increase [24,25].

Currently, progress has been made in the study of DFAA, but the following problems remain for further study: (1) Due to the lack of study on DFAA with a long sequence scale, there is no distinct definition of its temporal and spatial scales. (2) Previous studies focused on historical DFAA, and there is a lack of estimation of the evolution law of DFAA in the future. (3) Previous studies have focused on the evolution law analysis of DFAA, single cause analysis or studies on its law evolution and coping mechanisms but have not been connected with water pollution to form a systematic research method for DFAA. Therefore, this study attempted to estimate the change characteristics of future DFAA in the Hetao area by using the Representative Concentration Pathways (RCPs) climate scenario, and to estimate the occurrence probability of water pollution in future DFAA scenarios by simulating future water quality changes in Hetao area. Moreover, the characteristics of DFAA in this area were studied to cope with emergencies caused by climate change in advance, further provided a scientific basis for drought and flood control as well as water pollution control in the Hetao area.

2. Materials and Methods

2.1. Study Area

The Hetao area is located in the southern part of Bayannaoer City, Inner Mongolia Autonomous Region (Figure 1). The Hetao area has a complex spatial structure with an average elevation difference of less than 37 m. In addition, large-scale and long-term water resource development and other human activities in the irrigation area interfere with the canal system for irrigation and drainage. The annual precipitation is 50 to 250 mm, with large annual fluctuation. The precipitation in summer accounts for over 60% of the annual precipitation and that in spring only accounts for 10% to 20% [26].

The drought in spring is especially serious. Thus, DFAA typically occurs at the turn of spring to summer. The main river systems in the Hetao area are the Yellow River and Ulansuhai Nur Lake. Ulansuhai Nur Lake is the largest lake in the Hetao area, which is a part of the Yellow River [27]. Due to the large amount of agricultural wastewater discharge in the Hetao area, more than 90% of farmland irrigation (approximately 1.07×10^6 ha) water is discharged into Ulansuhai Nur Lake [27]. In addition, with the rapid development of local industries and population growth in recent years, industrial and domestic wastewater discharge has begun to exceed the standard [26]. As a result, the pollution load of TN and TP in the Ulansuhai Nur Lake area far exceeds the carrying capacity of the lake [28]. In summary, frequent DFAA in the Hetao area will lead to sudden water pollution in the Ulansuhai Nur Lake, thereby damaging the industrial and agricultural production and ecological environment in this region. Therefore, it is of both scientific and practical significance to study the characteristics and evolution of DFAA in the Hetao area and to further study the water quality of the Ulansuhai Nur Lake.

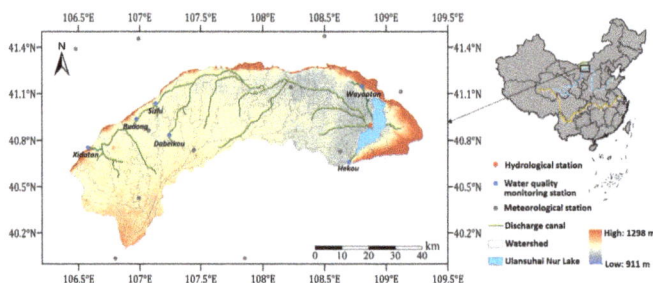

Figure 1. Location of the Hetao area.

2.2. Research Ideas

Precipitation and temperature are the main elements of climate change considered [3] in this paper. It is assumed that future point source emissions, withdrawn water, dam scheduling, cultivated land areas and irrigation systems, etc. are maintained at the current level. The assessment of the impacts of precipitation and temperature changes on DFAA in the Hetao area as follows (Figure 2):

(1) Changes in precipitation and temperature cause changes in the runoff process [9], thereby exerting direct impact on the temporal and spatial distribution and frequency of future DFAA. This section is based on the RCPs climate scenarios, aiming to estimate future precipitation and temperature in the Hetao area. LDFAI and SDFAI were calculated by using the future precipitation of the Hetao area.

(2) DFAA and its typical spatial distribution are prone to cause sudden major water pollution in the Ulansuhai Nur Lake [17,20]. We can predict the locations and time of potential DFAA in the future and respond to possible extreme DFAA in advance. This part of assessment used the established distributed water quantity and quality coupling model and future precipitation and temperature data to simulate and analyze the spatial and temporal changes in the water quantity and quality in the Ulansuhai Nur Lake inlet. The probability of joint distribution of DFAA and water quality in the lake inlet was constructed through the Copula function to estimate the probability of sudden water pollution in future DFAA scenarios.

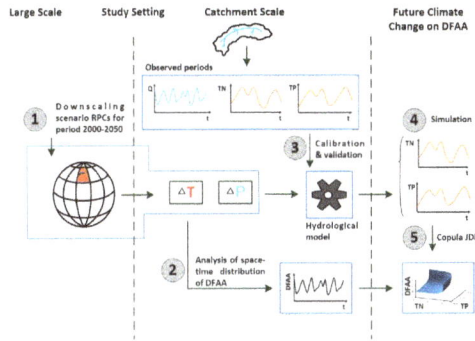

Figure 2. The framework of the impact evaluation of climate change on DFAA and water quality: (1) Selection of climate scenario and derivation of meteorological time series for future climate conditions, (2) Analysis of the spatial and temporal changes of DFAA, (3) Calibration and validation of a hydrological model, (4) Simulate TN and TP data from 1961 to 2050, (5) Calculate multivariate JPD of DFAA, TN and TP.

2.3. Evaluation Indicators

2.3.1. DFAA Indexes

To conduct a quantitative study on the scientific content and basic characteristics of summer LDFAI, researchers defined LDFAI as [19,20]:

$$\text{LDFAI} = (P_N - P_P) \times (|\ P_P\ | + |\ P_N\ |) \times 1.8^{-|P_P + P_N|} \tag{1}$$

where the time scale for calculating the LDFAI is defined as one year; P_P is the pre-flood standard precipitation; P_N is the post-flood standard precipitation; $(P_N - P_P)$ is the DFAA intensity term; $(|\ P_P\ | + |\ P_N\ |)$ is the drought and flood intensity term; $1.8^{-|P_P + P_N|}$ is the weight coefficient, of which the function is to increase the weight of long cycle DFAA and reduce the weight of pure drought or flood. The standard deviation of precipitation anomalies greater than 0.5 is considered flood, greater than 1 is significant flood, less than -0.5 is drought, less than -1 is significant drought; and between -0.5 and 0.5 is normal precipitation [17]. Based on the hydrometeorological characteristics of the middle and lower reaches of the Yangtze River, in this study the time scale for calculating the LDFAI is defined as two months, i.e., May and June (pre-flood period) and July and August (post-flood period). The long cycle DFAA is judged as LDFAI greater than 1 is a drought-to-flood incident, less than -1 is a flood-to-drought incident, and between -1 and 1 is normal. The greater the absolute value of LDFAI, the more serious the DFAA is.

The SDFAI [19,20] is essentially consistent with the LDFAI, which is expressed as:

$$\text{SDFAI} = (P_j - P_i) \times (|\ P_i\ | + |\ P_j\ |) \times 3.2^{-|P_i + P_j|} \tag{2}$$

where the time scale for calculating the SDFAI is defined as one month; P_i is the pre-flood standard precipitation; P_j is the post-flood standard precipitation. $3.2^{-|P_i + P_j|}$ is the weight coefficient; $j = i + 1(i = 5, 6, 7, 8, 9)$. The short cycle DFAA is judged as: SDFAI greater than 1 is a drought-to-flood incident, that less than -1 is a flood-to-drought incident and that between -1 and 1 is normal. The greater the absolute value of SDFAI, the more serious the DFAA incident is.

2.3.2. Evaluation of the Relationship between DFAA and Water Quality

Copula is a joint distribution function in a uniform distribution over the interval [0,1] [29,30]. Given F is an n-dimensional distribution function, and the edge distribution of each variable is

$F_1, F_2 \dots, F_n$, then there is an n-dimensional Copulas function C. For any $x \in R_n$, the distribution function satisfies:

$$F(x_1, x_2, \dots, x_n) = (X_1 \leq x_1, X_2 \leq x_2, \dots, X_n \leq x_n) = C[F_1(x_1), F_2(x_2), \dots, F_n(x_n)] \tag{3}$$

where $x_1, x_2, \dots, x_n$ are the observed samples, and $F(x)$ is the edge distribution function. Three Archimedean Copulas functions were selected, Gumbel-Hougaard (GH), Clayton and Frank, for joint distribution of two-dimensional and three-dimensional DFAA. The parameter estimation of the Copulas function was conducted by the appropriate line method, and the optimal Copula function was selected by a fit test to analyze the relationship between the DFAA and water quality.

With the constructed multivariate joint probability distribution (JPD) of DFAA, TN and TP, two DFAA situations under different water quality conditions in the Hetao area can be analysed:

(1) For future severe DFAA events, the joint transcendence probability is given more attention [31], and the joint probability analysis of TN, the most typical pollutant in the Hetao area, and DFAA, was selected. The multivariate JPD under this condition is denoted as $G_{X,Z}(x, z)$ and can be written as:

$$G_{X,Z}(x, z) = P_{X,Z}(X > x, Z > z) = 1 - F_X(x) - F_Z(z) + C(F_X(x), F_Z(z)) \tag{4}$$

$$G_{X,Z}(x, z) = P_{X,Z}(X > x, Z < z) = F_Z(z) - C(F_X(x), F_Z(z)) \tag{5}$$

where X denotes TN and x is its specific value, Z denotes DFAA and z is its specific value, $G_{X,Z}(x, z)$ is the bivariate JPD of pair (TN,DFAA), $F_X(x)$, $F_Z(z)$ are the marginal distribution function of TN, DFAA, respectively.

(2) When TN and TP exceed a specific value, DFAA is less than or more than a specific value, which addresses the water pollution risk under the DFAA condition. The multivariate JPD under this condition is denoted as $G_{X,Y,Z}(x, y, z)$ and presented as:

$$\begin{aligned} G_{X,Y,Z}(x, y, z) &= P_{X,Y,Z}(X > x, Y > y, Z > z) \\ &= 1 - F_X(x) - F_Y(y) - F_Z(z) + C(F_X(x), F_Y(y)) + C(F_X(x), F_Z(z)) \\ &+ C(F_Y(y), F_Z(z)) - C(F_X(x), F_Y(y), F_Z(z)) \end{aligned} \tag{6}$$

$$\begin{aligned} G_{X,Y,Z}(x, y, z) &= P_{X,Y,Z}(X > x, Y > y, Z < z) \\ &= F_Z(z) - C(F_X(x), F_Z(z)) - C(F_Y(y), F_Z(z)) + C(F_X(x), F_Y(y), F_Z(z)) \end{aligned} \tag{7}$$

where X denotes TN and x is its specific value; Y denotes TP and y is its specific value; Z denotes DFAA and z is its specific value; $G_{X,Y,Z}(x, y, z)$ represents the trivariate JPD of pair (TN, TP, DFAA); $F_X(x)$, $F_Y(y)$, and $F_Z(z)$ are the marginal distribution functions of TN, TP and DFAA, respectively.

2.4. Data Collection and Arrangement

The assessment of DFAA in the Hetao area requires data such as geographic information (GIS), monitoring data of meteorology, hydrology and environment [32]. GIS data include digital elevation models (DEM), water systems, vegetation maps, soil maps, meteorological sites, sewage outlets, runoff, and water quality sites distribution; the meteorological data include the sequence of meteorological elements such as historical daily precipitation and the maximum and minimum temperature of each site, and precipitation and temperature data under different meteorological elements of the Global Climate Model (GCM) in the future; the hydrological and water environment data include information such as monitoring section runoff, water quality concentration, and point source discharge. The basic data are shown in Table 1.

Table 1. The database of the climate change impact assessment in the Hetao area.

Type	Data	Scale	Source
GIS	DEM	Grid (90 m × 90 m)	Institute of Geographic Sciences and Natural Resources Research
	Land use	1:1,000,000	Institute of Geographic Sciences and Natural Resources Research
	Agrotype	1:4,000,000	Institute of Geographic Sciences and Natural Resources Research
Meteorology	Meteorological station	11 stations (1961–2017)	China Meteorological Administration
	GCM	Grid (1 × 1 °C) (2001–2050)	IPCC Fifth Assessment Report
Hydrology	Hydrological station	1 station (1980–2000)	Hetao Irrigation Administration Bureau
	Water quality station	6 stations (2012–2015)	Hetao Irrigation Administration Bureau

(1) The RCPs climate scenario

Due to space limitations, the RCPs climate scenario was selected to output precipitation and temperature series, and the 2001 to 2017 rate was periodically divided, and 2018 to 2050 was the forecast period. The precipitation and temperature were statistically analyzed and simulated in the RCP 2.6, RCP 4.5 and RCP 8.5 (as 1, 2 and 3) scenarios through the calibrated and down-scaled GFDL-ESM2M, HadGEM2-ES, IPSL-CM5A-LR, MIROC-ESM-CHEM, and NorESM1-M models (as a, b, c, d and e) from 2001 to 2017 [33]. If the data has uncertainty in the time of acquisition, or the time precision of the data is not met, then the validity of the spatio-temporal data cannot be explained, and reasonable reasoning cannot be made for the uncertain data in the data. Therefore, it is necessary check the uncertainty of the data. For the description of the difference between the RCPs climate scenario, the current mainstream international adopts the Taylor chart method, which is a way to integrate the three indicators of standard deviation (Std), root mean square error (RMSE) and correlation coefficient (R^2) into a concentrated display. The 15 climate models are represented by Taylor diagram and the results are shown in Figure 3. The results were compared with the measured values. HadGEM2-ES in the RCP 8.5 scenario had minimum uncertainty and relatively good precipitation and temperature results, and the correlation coefficients were 0.805 and 0.753, respectively. The results show that the climate model has the lowest uncertainty in all models, accurately simulating historical precipitation changes, and can be used for future precipitation prediction.

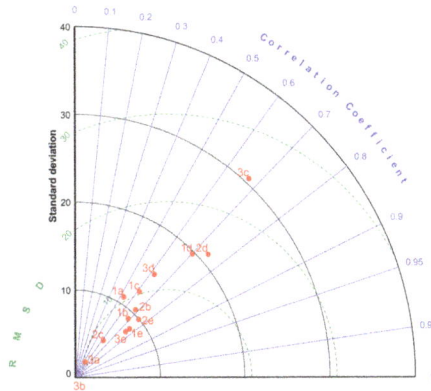

Figure 3. RPC 2.6, 4.5 and 8.4 scenario 15 models of Hetao area annual precipitation estimate Taylor diagram.

In the RCP 8.5 scenario, precipitation and temperature in the Hetao area showed a consistent increase, and compared with the period from 1961 to 2017, the RCP 8.5 scenario showed that precipitation increased by 12.2 mm and the temperature increased by 1.9 °C. When only considering the temperature increase, the frequency and intensity of the drought in the Hetao area will increase in the future; when merely considering the concentrated and increased precipitation, the runoff in the basin will increase, and the flood frequency will increase. The risk of drought and flood caused by climate change will directly affect the yield of local wheat and other crops. Therefore, it is necessary to study the law of DFAA in the area.

(2) Calibration and valibration of SWAT Model

The distributed coupling model of water quantity and water quality was developed by Dr. Jeff Arnold in 1998, that is, the watershed integrated management SWAT model [32]. The SWAT model can simulate the influence of climate change on runoff and water quality by changing the input data for scenario design and analysis [34,35]. Because of the large area of plain area, the traditional sub-basin division based on DEM has some difficulties. Based on this, the pre-defined tool of SWAT model is used to quantitatively simulate the water balance and water cycle characteristics of Hetao Irrigation Area under the influence of natural and human activities on the basis of artificially defined river course and sub-basin boundary. The SWAT model parameter calibration period of this study is from 1980 to 1997, and the inspection period is from 1980 to 1997 in this study, and the validation period is from 1998 to 2000. However, due to the limitation of water quality data and measured water quality data, only the monthly concentration data of six stations from 2012 to 2015 were selected to determine the relevant model water quality parameters. When the model structure and input parameters are preliminarily determined, it is necessary to calibrate and verify the model. In this paper, two indicators were selected to evaluate the applicability of the model, the model efficiency coefficient Ens proposed by Nash-Sutcliffe [36] and the correlation coefficient R^2 [34]. The Latin hypercube sampling and one-factor-at-a-time (LHS-OAT) method was used to analyze the sensitivity of the parameters. According to the model requirements, when $R^2 \geq 0.6$ and Ens ≥ 0.45, the simulation results are acceptable. The results of the runoff parameter rate are shown in Table 2. It can be seen from Table 2 that the SWAT model reached the basic evaluation criteria of R^2 and Ens for the simulation results of the total drainage flow and the simulation results of each water quality station. Therefore, the SWAT model is applicable to the simulation of flow and water quality in the Hetao area. The precipitation and temperature data from 1961 to 2050 were input into the SWAT model, which simulated the flow and water quality changes in future climate change scenarios, and then the impact of future climate change on DFAA can be studied.

Table 2. Evaluation of the simulation results of monthly TN and TP during calibration and validation periods.

Station	Variety	Calibration		Validation	
		R^2	Ens	R^2	Enst
Zongpaigan	Runoff	0.69	0.61	0.73	0.63
Xidatan	TN	0.8	0.74	0.46	0.45
	TP	0.69	0.68	0.62	0.51
Wayaotan	TN	0.79	0.77	0.79	0.7
	TP	0.76	0.74	0.81	0.67
Budong	TN	0.72	0.67	0.89	0.46
	TP	0.68	0.57	0.73	0.57

Table 2. *Cont.*

Station	Variety	Calibration		Validation	
		R^2	Ens	R^2	Enst
Dabeikou	TN	0.71	0.56	0.75	0.51
	TP	0.81	0.56	0.62	0.52
Hekou	TN	0.62	0.51	0.71	0.48
	TP	0.73	0.53	0.67	0.58
Sizhi	TN	0.73	0.61	0.63	0.52
	TP	0.62	0.49	0.65	0.54

3. Results

3.1. Law Analysis of DFAA

3.1.1. Analysis of DFAA on the Time Scale

On the time scale, we compared the trend of the long-term sequence DFAA rather than the characteristics in changes within a year. Therefore, we selected LDFAI for the time-scale analysis of DFAA. The precipitation from 1961 to 2050 was standardized as Equation (1). As shown in Figure 4a, there may be a multiple-level time scale structure and localization characteristics of changes in DFAA in the time domain. Wavelet analysis with the time-frequency multi-resolution function proposed by Morlet provides the possibility of effectively studying the problems of time series, which can clearly reveal multiple change cycles hidden in the time series [37].

The Morlet wavelet analysis of the LDFAI from 1961 to 2050 was conducted to reveal the DFAA cycle in the Hetao area over the past 57 years and predict the change law from 2018 to 2050. The wavelet variance in Figure 4b can reflect the distribution of the wave energy of the LDFAI with time scale, and determine the main cycle during DFAA. There are four distinct peaks in Figure 4b, 3, 6, 11 and 21a, indicating that the quasi-cycles of these four scales play a major role in the DFAA in the Hetao area [37]. The largest peak corresponds to the characteristic time scale of 21a, indicating that it has the strongest periodic turbulence in the most significant cycle, which is the first main cycle of drought and flood changes in the area. The second, third and fourth main cycles of drought and flood changes are 11, 6 and 3a, respectively, representing the following: 21a characteristic time scale has approximately 7 cycles of drought and flood alternation; the average cycle of drought and flood changes is approximately 13a; the drought and flood centers are in 1966, 1980, 1997, 2010, 2022, 2035, 2050; on the 25a characteristic time scale, there are approximately 4 cycles of drought and flood alternation; the average change cycle is approximately 16a; on the 11a characteristic time scale, there are approximately 10 cycles of drought and flood alternation; and the average change cycle is approximately 9a. Regardless of the time scale, the square of modulus of the wavelet coefficient after 2017 has shown a weakening trend, indicating that DFAAs in the Hetao area will decrease in the future. The red (blue) rendering in Figure 4c refers to that where the real part of the wavelet coefficient is positive (negative), and the darker the color is, the greater the degree of the drought or flood. The square of modulus of the wavelet coefficient in Figure 4d refers to the wavelet energy spectrum, and the larger the value (the darker the red), the stronger the wave energy is, and the more significant the cycle is. The two most energy-concentrated centers in Figure 4d represent the characteristics of the changes in wave energy. They are (1) scale of 22 to 26a, wave energy is from the late 1970s to early 1990s, and (2) scale of 19a, the wave energy is the strongest and runs through 2000 to 2025, with the strongest performance.

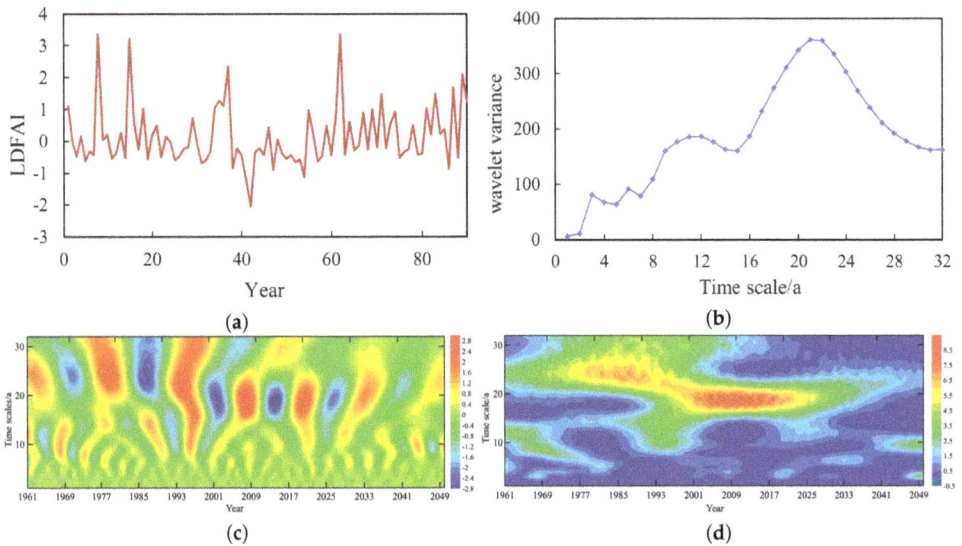

Figure 4. Morlet wavelet analysis of DFAA in the Hetao area during 1961 to 2050. (a) LDFAI; (b) Wavelet variance map; (c) Real part of wavelet transform coefficients; (d) Square of modulus of wavelet transform coefficients.

3.1.2. Analysis of Spatial Differentiation Characteristics of DFAA

On the spatial scale, we pay attention to the spatial position of DFAA occurrence in different months. Therefore, we select SDFAI for the spatial-scale analysis of DFAA (Equation (2)). The spatial distribution of the frequency of DFAA in 11 hydrological stations of the Hetao area from May to July and from July to September are shown in Figure 5. Generally, the precipitation in the Hetao area from July to September is greater than that from May to July, thus the frequency of DFAA in the Hetao area is also reversed; from May to July, the drought-to-flood incidents are frequent, and from July to September, the flood-to-drought incidents are frequent. The spatial distribution of the frequency of DFAA in the Hetao area is uneven, and generally, there is no consistency between drought-to-flood incidents and flood-to-drought incidents in the same period. That is, it is difficult to form short-cycle drought-to-flood-to-drought and flood-to-drought-to-flood incidents in the Hetao area. During the 57 years from 1961 to 2017, the high-frequency DFAA region in the Hetao area is generally concentrated in the east, thereby Ulansuhai Nur Lake has become the most frequent DFAA area, which is unfavourable for maintaining the ecological stability of the lake. However, there are few DFAA in the northern part of the Hetao area. The reason is that the precipitation in the area is relatively balanced from April to September, so DFAA is not frequent. However, from 2018 to 2050, the SDFAI in the Hetao area showed a significant decrease compared with the previous period, which is completely consistent with the previous results of the LDFAI, and the frequent occurrence area has shifted westward.

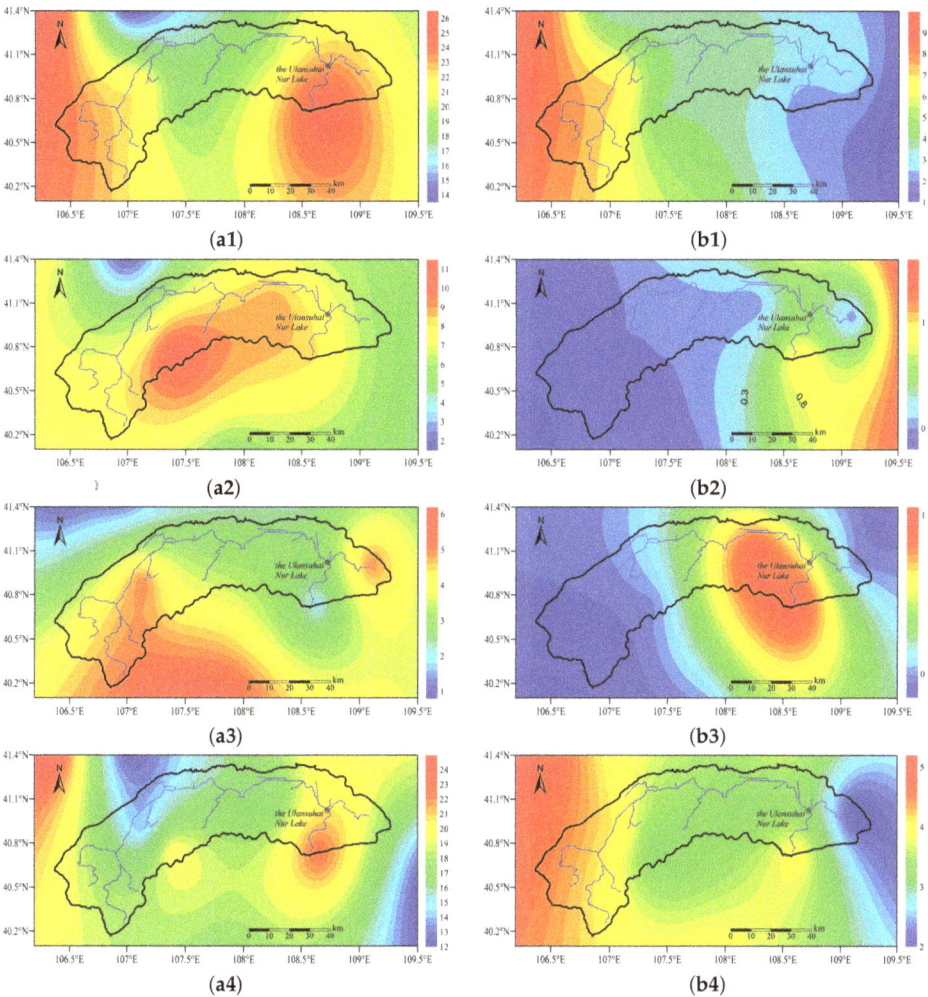

Figure 5. Spatial distribution of the frequency of flood-drought abrupt alternation events in the Hetao Area. (**a1**) Flood-to-drought from May to July in 1961–2017; (**b1**) Flood-to-drought from May to July in 2018–2050; (**a2**) Drought-to-flood from May to July in 1961–2017; (**b2**) Drought-to-flood from May to July in 2018–2050; (**a3**) Flood-to-drought from July to September in 1961–2017; (**b3**) Flood-to-drought from July to September in 2018–2050; (**a4**) Drought-to-flood from July to September in 1961–2017; (**b4**) Drought-to-flood from July to September in 2018–2050.

3.2. The Relationship between DFAA and Water Quality

The contour plots of different periods of $G_{X,Z}(x,z)$ are given in Figure 6 (Equation (4) and (5)). They represent the JPD of changes in water quality in past and future natural precipitation conditions, so the joint probability with various given combinations of pair (TN, SDFAI) with a given certain joint probability can be obtained [29–31]. It can be seen from Figure 6 that the value of JPD which exceeds a certain probability becomes larger with more TN and high SDFAI. When $G_{X,Z}(x,z)$ is between 0.1 to 0.3 or 0.8 to 0.9, the contour plot spacing is larger and TN and SDFAI change more obviously compared with that between 0.3 and 0.8. It can be seen from Figure 6 that when the joint distribution probability is determined, we can also compare the TN overshoot probability corresponding to different SDFAI values. For example, when the joint distribution probability is 0.1 and the SDFAI value exceeds 1.5, the TN value corresponding to Figure 6a is greater than 1.92 mg/L, and the corresponding TN value of Figure 6b is greater than 2.11 mg/L. This result indicates that in future climate change scenarios, when DFAA occurs sharply, the probability of water quality exceeding the standard will increase. The possibility of constructing different combinations of events through conditional probability is listed in Table 3. Between 1961 and 2017, when a drought-to-flood event occurs (SDFAI > 1), the probability of TN pollutants exceeding Class III is 0.608; when a flood-to-drought occurs (SDFAI < −1), the probability of TN pollutants exceeding Class III is 0.732; when there is no flood-to-drought or drought-to-flood event (−1 ≤ SDFAI ≤ 1), the probability of TN pollutants exceeding Class III is 0.595. Between 2018 and 2050, when a drought-to-flood event occurs (SDFAI > 1), the probability of TN pollutants exceeding Class III is 0.712; when a flood-to-drought event occurs (SDFAI < −1), the probability of TN pollutants exceeding Class III is 0.797; when there is no flood-to-drought or drought-to-flood event (−1 ≤ SDFAI ≤ 1), the probability of TN pollutants exceeding Class III is 0.432. This result indicates that when DFAA occurs, the probability of TN pollutants in the Ulansuhai Nur Lake entrance is lower than usual, and the probability of TN pollutants accompanying DFAA will increase in the future.

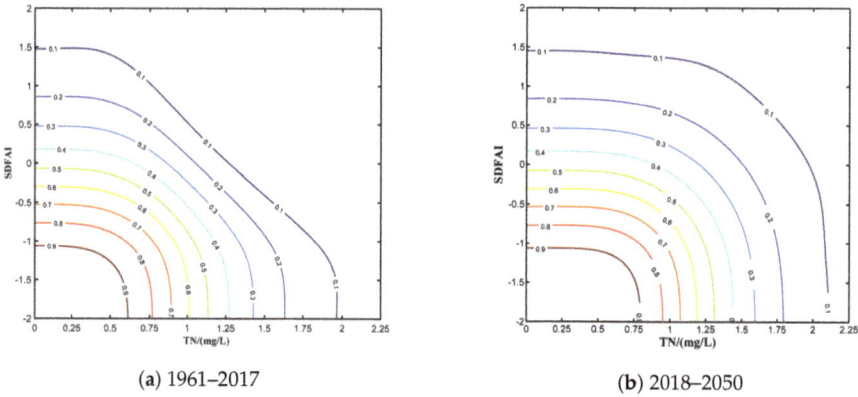

(a) 1961–2017   (b) 2018–2050

Figure 6. The contour plot of bivariate JPD of TN and SDFAI.

Table 3. The possibility of different bivariate combination events.

Year	P (TN > 1 | SDFAI > 1)	P (TN > 1 | SDFAI < −1)	P (TN > 1 | −1 ≤ SDFAI ≤ 1)
1961–2017	0.576	0.732	0.578
2018–2050	0.712	0.797	0.432

The contour surface of $G_{X,Y,Z}(x,y,z)$ is shown in Figure 7 (Equations (6) and (7)), which indicates that close dependent correlation exists in TN, TP and SDFAI. Figure 7 expounds the JPD of changes in water quality in past and future natural precipitation conditions, thus the joint probability with various given combinations of pair (TN, TP, SDFAI) as well as the various combinations of pair (TN, TP, SDFAI) with a given certain joint probability can be achieved [29–31]. It can be seen from Figure 7 that the value of JPD which does not exceed a certain probability tends to be large with small TN and TP and low SDFAI. When $G_{X,Y,Z}(x,y,z)$ is between 0.7 to 0.9, the contour surface spacing is larger and TN and TP change more obviously compared with that between 0.1 and 0.7. Similarly, different three-variable event combinations are constructed by conditional probabilities, as shown in Table 4. Between 1961 and 2017, when a drought-to-flood event occurs (SDFAI > 1), the probability of one pollutant exceeding the standard or two pollutants exceeding the standard at the same time is 0.939; when a flood-to-drought event occurs (SDFAI < −1), the probability of one pollutant exceeding the standard or both pollutants exceeding the standard at the same time is 0.827; when there is no flood-to-drought or drought-to-flood event ($-1 \leq$ SDFAI ≤ 1), the probability is 0.346. Between 2018 and 2050, when a drought-to-flood event occurs (SDFAI > 1), the probability of one pollutant exceeding the standard or two pollutants exceeding the standard at the same time is 0.956; when a flood-to-drought event occurs (SDFAI < -1), the probability of one pollutant exceeding the standard or both pollutants exceeding the standard at the same time is 0.851; when there is no flood-to-drought or drought-to-flood event ($-1 \leq$ SDFAI ≤ 1), the probability is 0.336. This result further confirms the results of the bivariate joint distribution. Moreover, the probability of water quality exceeding the standard caused by drought-to-flood events in the joint distribution of three variables is greater than that in flood-to-drought events. Therefore, DFAA and the risk of water pollution in the Hetao area can be clarified more reasonably by using trivariate JPD, and it also provides technological guidance for irrigation planning and DFAA resistance.

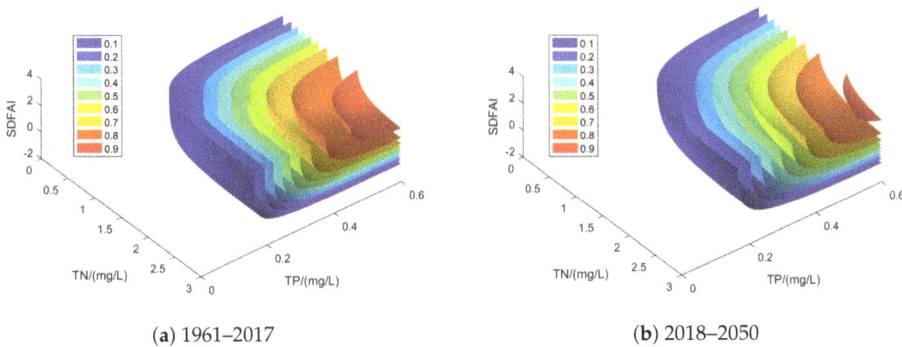

(a) 1961–2017   (b) 2018–2050

Figure 7. Joint transcendental probability section diagram.

Table 4. The possibility of different trivariate combination events.

Year	P (A* | SDFAI > 1)	P (A* | SDFAI < −1)	P (B* | −1 ≤ SDFAI≤ 1)
1961–2017	0.94	0.827	0.346
2018–2050	0.958	0.851	0.336

A^*: One pollutant exceeds the standard or two pollutants exceeded the standard simultaneously; B^*: Two pollutants do not exceed the standard simultaneously.

4. Discussion

The increase in the evaporation rate caused by climate warming resulted in the decrease in available water and the increase in drought events in the mid-latitudes and semi-arid low latitudes [38]. In the future, the increase in temperature and extreme precipitation will aggravate water pollution and accelerate the occurrence of eutrophication events [4,5,39]. These results can also be found in the IPCC report [40]. We believe that under the premise of the reduction in DFAA events in the Hetao area, there are three reasons for the increasing water pollution risk:

(1) With the increase in precipitation processes, rainwater and runoff pass through the ground, and the pollutants accumulated on the surface are carried into the water body, causing pollution of surface water and even groundwater within the drainage area, especially in the vicinity of farmland or industrial land, which will form serious non-point source pollution [41]. Therefore, the change in precipitation intensity and frequency will affect non-point source pollution. As two of the main elements of non-point source pollution, nitrogen and phosphorus are greatly affected by precipitation process. If precipitation and its strength increase, then the runoff scouring effect will intensify, and the nitrogen and phosphorus loads flowing into the water body will increase accordingly [42].

(2) With the increase in air temperature, the water surface temperature will also increase, which leads to an increase in the temperature difference and thermocline in the upper and lower layers of water. The presence of thermoclines can lead to the formation of anoxic layers at the bottom of water bodies such as rivers or lakes. Nitrogen and phosphorus release easily from sediment to bottom water in an anoxic bottom water environment, and lead to an increase in nitrogen and phosphorus concentrations in surface water, which is the main reason that nitrogen and phosphorus loads increase with surface runoff coming into the water environment. The increase in water temperature will also increase the activities of microorganisms and promote the release of endogenous nitrogen and phosphorus in sediment. If the nitrogen concentration in the water reaches a certain level, eutrophication will be intensified when environmental conditions such as temperature and light are satisfied [43].

(3) Under drought conditions, runoff is reduced and the water temperature is relatively high, which will increase the concentration of NH_4^+ and NO_2^- in the water. Some studies have shown that concentrations of NH_4^+ and NO_2^- increased by 1.9 and 1.3 times, respectively, in a dry year and a normal year [44]. The increasing frequency of DFAA will cause a large number of surface pollutants to enter water bodies. Drought-to-flood incidents were taken as an example: in the early stage of drought, the flow rate of the river channel decreased, leading to a decrease in the ability to dilute and transport substances and an increase in the concentration of pollutants in the water body and surrounding farmlands. In the later stage of rapid formation of flood, the hydrodynamic conditions increased rapidly, directly bringing a multitude of pollutants in the surrounding farmlands and the river channel into the Ulansuhai Nur Lake. These processes may occur simultaneously with DFAA. At the same time, DFAA will also cause a large amount of sediment to enter water bodies or cause sediment resuspension, which will affect the sediment content of the water body, thus further affecting the transport and transformation of pollutants, and water quality [45,46].

5. Conclusions

According to daily precipitation data from 1961 to 2050 in the Hetao area, monthly TN and TP data, and LDFAI and SDFAI analysis, this paper comprehensively analyzed changes in the DFAA trend in the Hetao area, the spatial and temporal distribution characteristics and the risk of water pollution caused by DFAA. The main conclusions are as follows:

(1) In the Hetao area, the phenomenon of the LDFAI was mainly for drought-to-flood events, and there is a trend that the frequency of DFAA will decrease in the future; the drought-to-flood incidents occurred frequently in May to July among the SDFAI, and flood-to-drought incidents occurred frequently from July to September.

(2) Due to the uneven distribution of precipitation in the flood season in the Hetao area, the spatial distribution of the DFAA is not uniform; during the 57 years from 1961 to 2017, the high-frequency DFAA regions in the Hetao area were generally concentrated in the Ulansuhai Nur Lake in the eastern part of the region. From 2018 to 2050, frequent occurrences of DFAA occurred in the west.

(3) The Copula function is used to calculate the JDP of SDFAI, TN and TP. The risk of water quality exceeding the standard will increase when the DFAA happens, and the probability of water quality exceeding the standard caused by drought-to-flood in the three variable joint distribution is greater than that in flood-to-drought.

(4) Extreme weather such as an increase in future temperatures and an increase in extreme precipitation will exacerbate water pollution, causing further increases in the risk of excessive water quality in future DFAA, which is consistent with the conclusions of the IPCC report. The results can provide a basis for flood control and drought resistance and pollution control in the Hetao area.

Water **2019**, *11*, 652

Author Contributions: This study was designed by Y.Y. and B.W., W.B. and T.X. contributed to data collection and calculation. Y.Y. carried out calculation analysis and completed the manuscript. D.Y. and J.M. contributed to the manuscript modification.

Funding: This research was funded by the National Key Research and Development Project (No. 2016YF - A0601503), the National Natural Science Foundation of China (No. 91547209), the National Key Research and Development Project (No. 2017YFA0605004), and the Public Service Special Project of the Environmental Protection Ministry of China (No. 201509027).

Conflicts of Interest: The authors declare no conflict of interest.

Abbreviations

The following abbreviations are used in this manuscript:

DFAA	Drought-flood abrupt alternation
RCPs	Representative Concentration Pathways
TN	Total nitrogen
TP	Total phosphorus
IPCC	Intergovernmental Panel on Climate Change
DEM	Digital Elevation Model
GCM	Global Climate Model
LDFAI	Long-cycle drought-flood abrupt alternation index
SDFAI	Short-cycle drought-flood abrupt alternation index
JPD	Joint probability distribution

References

1. IPCC. Summary for Policymakers. In *Global Warming of 1.5 °C; An IPCC Special Report on the Impacts of Global Warming of 1.5 °C above Pre-Industrial Levels and Related Global Greenhouse Gas Emission Pathways, In the Context of Strengthening the Global Response to the Threat of Climate Change, Sustainable Development, and Efforts to Eradicate Poverty*; Intergovernmental Panel on Climate Change (IPCC): Geneva, Switzerland, 2018.
2. Walther, G.R.; Post, E.; Convey, P.; Menzel, A.; Parmesan, C.; Beebee, T.J.C.; Fromentin, J.M.; Hoegh-Guldberg, O.; Bairlein, F. Ecological responses to recent climate change. *Nature* **2002**, *416*, 389–395. [CrossRef] [PubMed]
3. Rogelj, J.; Meinshausen, M.; Knutti, R. Global warming under old an new scenarios using IPCC climate sensitivity range estimates. *Nat. Clim. Chang.* **2012**, *2*, 248–253. [CrossRef]
4. Harvell, C.D.; Mitchell, C.E.; Ward, J.R.; Altizer, S.; Dobson, A.P.; Ostfeld, R.S.; Samuel, M.D. Climate warming and disease risks for terrestrial and marine biota. *Science* **2002**, *296*, 58–62. [CrossRef] [PubMed]
5. Walter, K.M.; Zimov, S.A.; Chanton, J.P.; Verbyla, D.; Chapin, F.S. Methane bubbling from Siberian thaw lakes as a positive feedback to climate warming. *Nature* **2006**, *443*, 71–75. [CrossRef] [PubMed]
6. Parmesan, C.; Yohe, G. A globally coherent fingerprint of climate change impacts across natural systems. *Nature* **2003**, *443*, 37–42. [CrossRef]
7. Yang, S.Y.; Wu, B.Y.; Zhang, R.H.; Zhou, S.W. Relationship between an abrupt drought-flood transition over mid-low reaches of the Yangtze River in 2011 and the intraseasonal oscillation over mid-high latitudes of East Asia. *Acta Meteorol. Sin.* **2013**, *27*, 129–143. [CrossRef]
8. Espinoza, J.C.; Ronchail, J.; Guyot, J.L.; Junquas, C.; Drapeau, G.; Martinez, J.M.; Santini, W.; Vauchel, P.; Lavado, W.; Ordonez, J.; et al. From drought to flooding: Understanding the abrupt 2010–11 hydrological annual cycle in the Amazonas River and tributaries. *Environ. Res. Lett.* **2012**, *7*, 1–7. [CrossRef]
9. Vogel, J.L.; Huff, F.A. Relation Between the St-Louis Urban Precipitation Anomaly and Synoptic Weather Factors. *J. Appl. Meteorol. Clim.* **1978**, *17*, 1141–1152. [CrossRef]
10. Trenberth, K.E.; Guillemot, C.J. Physical Processes Involved in the 1988 Drought and 1993 Floods in North America. *J. Clim.* **1996**, *9*, 1288–1298. [CrossRef]
11. Cook, B.I.; Seager, R.; Miller, R.L. Atmospheric circulation anomalies during two persistent north american droughts: 1932–1939 and 1948–1957. *Clim. Dyn.* **2011**, *36*, 2339–2355. [CrossRef]

12. Garnett, E.R.; Khandekar, M.L. The impact of large-scale atmospheric circulations and anomalies on Indian monsoon droughts and floods and on world grain yields-a statistical analysis. *Agric. For. Meteorol.* **1992**, *61*, 113–128. [CrossRef]

13. Hastenrath, S.; Polzin, D.; Mutai, C. Diagnosing the droughts and floods in equatorial East Africa during boreal autumn 2005–08. *J. Clim.* **2010**, *23*, 813–817. [CrossRef]

14. Zhang, Z.Z.; Chao, B.F.; Chen, J.L.; Wilson, C.R. Terrestrial water storage anomalies of Yangtze River Basin droughts observed by GRACE and connections with ENSO. *Glob. Planet. Chang.* **2015**, *126*, 35–45. [CrossRef]

15. Keyantash, J.; Dracup, J.A. The Quantification of Drought: An Evaluation of Drought Indices. *Am. Meteorol. Soc.* **2002**, *83*, 1167–1180. [CrossRef]

16. Tote, C.; Patricio, D.; Boogaard, H.; van der Wijngaart, R.; Tarnavsky, E.; Funk, C. Evaluation of Satellite Rainfall Estimates for Drought and Flood Monitoring in Mozambique. *Remote Sens.* **2015**, *7*, 1758–1776. [CrossRef]

17. Li, X.H.; Zhang, Q.; Zhang, D.; Ye, X.C. Investigation of the drought-flood abrupt alternation of streamflow in Poyang Lake catchment during the last 50 years. *Hydrol. Res.* **2017**, *48*, 1402–1417. [CrossRef]

18. Gao, C.; Tian, R. The influence of climate change and human activities on runoff in the middle reaches of the Huaihe River Basin, China. *J. Geogr. Sci.* **2018**, *28*, 79–92. [CrossRef]

19. Wu, Z.W.; Li, J.P.; He, J.H.; Jiang, Z.H. Occurrence of droughts and floods during the normal summer monsoons in the mid and lower reaches of the Yangtze River. *Geophys. Res. Lett.* **2006**, *33*, L05813. [CrossRef]

20. Wu, Z.W.; Li, J.P.; He, J.H.; Jiang, Z.H. Large-scale atmospheric singularities and summer long-cycle droughts-floods abrupt alternation in the middle and lower reaches of the Yangtze River. *Chin. Sci. Bull.* **2006**, *51*, 2028–2034. [CrossRef]

21. Mosavi, A.; Ozturk, P.; Chau, K.W. Flood Prediction Using Machine Learning Models: Literature Review. *Water* **2018**, *10*, 1536. [CrossRef]

22. Mosavi, A.; Edalatifar, M. A Hybrid Neuro-Fuzzy Algorithm for Prediction of Reference Evapotranspiration. In Proceedings of the International Conference on Global Research and Education, Iasi, Romania, 25–28 September 2017; pp. 235–243.

23. Choubin, B.; Moradi, E.; Golshan, M.; Adamowski, J.; Sajedi-Hosseini, F.; Mosavi, A. An ensemble prediction of flood susceptibility using multivariate discriminant analysis, classification and regression trees, and support vector machines. *Sci. Total Environ.* **2019**, *10*, 2087–2096. [CrossRef] [PubMed]

24. Yang, C.G.; Yu, Z.B.; Hao, Z.C.; Zhang, J.Y.; Zhu, J.T. Impact of climate change on flood and drought events in Huaihe River Basin, China. *Hydrol. Res.* **2012**, *43*, 14–22. [CrossRef]

25. Hurlbert, M.; Gupta, J. Adaptive Governance, Uncertainty, and Risk: Policy Framing and Responses to Climate Change, Drought, and Flood. *Risk Anal.* **2016**, *36*, 339–356. [CrossRef]

26. Zhu, D.N.; Ryan, M.C.; Sun, B.; Li, C.Y. The influence of irrigation and Wuliangsuhai Lake on groundwater quality in eastern Hetao Basin, Inner Mongolia, China. *Hydrogeol. J.* **2014**, *22*, 1101–1114. [CrossRef]

27. Wu, Y.; Shi, X.H.; Li, C.Y.; Zhao, S.N.; Pen, F.; Green, T.R. Simulation of Hydrology and Nutrient Transport in the Hetao Irrigation District, Inner Mongolia, China. *Water* **2017**, *9*, 169. [CrossRef]

28. Zhang, H.; Ma, D.; Hu, X. Arsenic pollution in groundwater from Hetao Area, China. *Environ. Geol.* **2002**, *41*, 638–643. [CrossRef]

29. Mou, S.Y.; Shi, P.; Qu, S.M.; Ji, X.M.; Zhao, L.L.; Feng, Y.; Chen, C.; Dong, F.C. Uncertainty Analysis of Two Copula-Based Conditional Regional Design Flood Composition Methods: A Case Study of Huai River, China. *Water* **2018**, *10*, 1872. [CrossRef]

30. Aranda, J.A.; Garcia-Bartual, R. Synthetic Hydrographs Generation Downstream of a River Junction Using a Copula Approach for Hydrological Risk Assessment in Large Dams. *Water* **2018**, *10*, 1570. [CrossRef]

31. Cai, W.Y.; Di, H.; Liu, X.P. Estimation of the Spatial Suitability of Winter Tourism Destinations Based on Copula Functions. *Int. J. Environ. Res. Public Health* **2019**, *16*, 186. [CrossRef]

32. Arnold, J.G.; Moriasi, D.N.; Gassman, P.W.; Abbaspour, K.C.; White, M.J.; Srinivasan, R.; Santhi, C.; Harmel, R.D.; van Griensven, A.; Van Liew, M.W.; et al. Swat: Model Use, Calibration, and Validation. *Trans. Asabe* **2012**, *55*, 1491–1508. [CrossRef]

33. Nazari-Sharabian, M.; Taheriyoun, M.; Ahmad, S.; Karakouzian, M.; Ahmadi, A. Water Quality Modeling of Mahabad Dam Watershed—Reservoir System under Climate Change Conditions, Using SWAT and System Dynamics. *Water* **2019**, *11*, 394. [CrossRef]

34. Abbaspour, K.C.; Yang, J.; Maximov, I.; Siber, R.; Bogner, K.; Mieleitner, J.; Zobrist, J.; Srinivasan, R. Modelling hydrology and water quality in the pre-alpine/alpine Thur watershed using SWAT. *J. Hydrol.* **2007**, *333*, 413–430. [CrossRef]

35. Srinivasan, R.; Ramanarayanan, T.S.; Arnold, J.G.; Bednarz, S.T. Large area hydrologic modeling and assessment—Part II: Model application. *J. Am. Water Resour. Assoc.* **19982008**, *34*, 142–149. [CrossRef]

36. Moriasi, D.N.; Arnold, J.G.; Van Liew, M.W.; Bingner, R.L.; Harmel, R.D.; Veith, T.L. Model Evaluation Guidelines for Systematic Quantification of Accuracy in Watershed Simulations. *Trans. Asabe* **2007**, *50*, 885–900. [CrossRef]

37. Kovacs, J.; Gaborhatvani, I.; Korponai, J.; Kovacs, I.S. Morlet wavelet and autocorrelation analysis of long-term data series of the Kis-Balaton water protection system (KBWPS). *Ecol. Eng.* **2010**, *36*, 1469–1477. [CrossRef]

38. Whitehead, P.G.; Wilby, R.L.; Battarbee, R.W.; Kernan, M.; Wade, A.J. A review of the potential impacts of climate change on surface water quality. *Hydrolog. Sci. J.* **2009**, *54*, 101–123. [CrossRef]

39. Ducharne, A. Importance of stream temperature to climate change impact on water quality. *Hydrol. Earth Syst. Sci.* **2008**, *12*, 797–810. [CrossRef]

40. Stocker, T.F.; Qin, D.; Plattner, G.K.; Tignor, M.M.B.; Allen, S.K.; Boschung, J.; Nauels, A.; Xia, Y.; Bex, V.; Midgley, P.M. *Climate Change 2013: The Physical Science Basis*; University Press: Cambridge, UK, 2013; pp. 1–33.

41. Kaste, O.; Wright, R.F.; Barkved, L.J.; Bjerkeng, B.; Engen-Skaugen, T.; Magnusson, J.; Saelthun, N.R. Linked models to assess the impacts of climate change on nitrogen in a Norwegian river basin and FJORD system. *Sci. Total Environ.* **2006**, *365*, 1–3. [CrossRef]

42. Whitehead, P.G.; Butterfield, D.; Wade, A.J. The title of the cited article. *Hydrol. Res.* **2009**, *40*, 113–122. [CrossRef]

43. Luo, Y.Z.; Ficklin, D.L.; Liu, X.M.; Zhang, M.H. Assessment of climate change impacts on hydrology and water quality with a watershed modeling approach. *Sci. Total Environ.* **2013**, *450*, 72–82. [CrossRef]

44. Vliet, M.T.H.V.; Zwolsman, J.J.G. Impact of summer droughts on the water quality of the meuse river. *J. Hydrol.* **2008**, *353*, 1–17. [CrossRef]

45. Breivik, K.; Wania, F.; Muir, D.C.G.; Alaee, M.; Backus, S.; Pacepavicius, G. Empirical and modeling evidence of the long-range atmospheric transport of decabromodiphenyl ether. *Environ. Sci. Technol.* **2006**, *40*, 4612–4618. [CrossRef] [PubMed]

46. Hilscherova, K.; Dusek, L.; Kubik, V.; Cupr, P.; Hofman, J.; Klanova, J.; Holoubek, I. Redistribution of organic pollutants in river sediments and alluvial soils related to major floods. *J. Soil Sediment.* **2007**, *7*, 167–177. [CrossRef]

water

MDPI

Article

Extreme Precipitation Spatial Analog: In Search of an Alternative Approach for Future Extreme Precipitation in Urban Hydrological Studies

Ariel Kexuan Wang [1,*], Francina Dominguez [2] and Arthur Robert Schmidt [1]

[1]   Department of Civil and Environmental Engineering, University of Illinois at Urbana-Champaign, Urbana, IL 61801, USA; aschmidt@illinois.edu
[2]   Department of Atmospheric Sciences, University of Illinois at Urbana-Champaign, Urbana, IL 61801, USA; francina@illinois.edu
*   Correspondence: wangkx2013@gmail.com

Received: 20 March 2019; Accepted: 10 May 2019; Published: 17 May 2019

Abstract: In this paper, extreme precipitation spatial analog is examined as an alternative method to adapt extreme precipitation projections for use in urban hydrological studies. The idea for this method is that real climate records from some cities can serve as "analogs" that behave like potential future precipitation for other locations at small spatio-temporal scales. Extreme precipitation frequency quantiles of a 3.16 km^2 catchment in the Chicago area, computed using simulations from North American Regional Climate Change Assessment Program (NARCCAP) Regional Climate Models (RCMs) with L-moment method, were compared to National Oceanic and Atmospheric Administration (NOAA) Atlas 14 (NA14) quantiles at other cities. Variances in raw NARCCAP historical quantiles from different combinations of RCMs, General Circulation Models (GCMs), and remapping methods are much larger than those in NA14. The performance for NARCCAP quantiles tend to depend more on the RCMs than the GCMs, especially at durations less than 24-h. The uncertainties in bias-corrected future quantiles of NARCCAP are still large compared to those of NA14, and increase with rainfall duration. Results show that future 3-h and 30-day rainfall in Chicago will be similar to historical rainfall from Memphis, TN and Springfield, IL, respectively. This indicates that the spatial analog is potentially useful, but highlights the fact that the analogs may depend on the duration of the rainfall of interest.

Keywords: spatial analog; extreme precipitation; future precipitation at urban scale; RCM uncertainty

1. Introduction

The population in urban areas has increased by over 250% since 1964, and as a result, urban hydraulic infrastructure has dramatically increased [1,2]. Due to increased impervious surfaces, expanded coverage of storm sewers and other engineered drainage systems, and modified landscape, urban hydrologic systems have much smaller storage and shorter response time than the pre-development hydrologic systems. As a consequence, high-intensity short-duration precipitation extremes become one of the major causes of failures of hydraulic infrastructure in urban hydrologic systems [3], and thus play a significant part in studies and designs of urban hydraulic infrastructures. Design storms in particular, which are precipitation events of a specified duration and return period, are used universally in design of storm sewers [4], storage facilities [5], bridges and culverts [3], and in floodplain studies [6]. The return periods of these design storms can range from 2-year [4] to 100-year [5,6], while the durations range from less than an hour to several days depending on the purpose and scope of the structure being designed. The widely accepted and applied way to calculate design storms is to conduct frequency analysis on observed rainfall data with a stationarity assumption.

However, as changes in extreme precipitation events have been observed [7] and are projected to continue [8] in a warmer climate by increasing studies, design storms analyzed using past data may not be applicable in the future [9]. Taking the state of Illinois as an example, flood damage on the order of billions of dollars was documented as a result from failures of urban hydraulic infrastructure under extreme rainfalls from 2007 to 2014 [4]. Urgent action is required in estimating future extreme rainfall to prevent losses and damages [10,11].

Even though extreme precipitation results from a combination of factors from large to local scales, global warming caused by anthropogenic greenhouse gases emission is likely to lead to the increase of precipitation extremes [8,12]. Atmospheric water holding capacity has increased [7,8] and is projected to increase in a warmer climate [8]. This increase in atmospheric humidity leads to increased frequency and intensity of the precipitation extremes, which are primarily controlled by variations in the atmosphere's moisture content [7,8,13–16] . This is consistent with current observations and climate model simulations [7,8,17]. Kunkel [17] found that the frequencies of extreme precipitation events increased in the latter portion of the 20th century in the U.S., and the timing of this increase matched with a fast increase in global average temperatures. Trenberth et al. [7] showed that the precipitation extremes are in general observed to have an increasing trend, even though the magnitude varies with the location, duration, and frequency of the precipitation extremes.

General Circulation Models (GCMs) are the most recognized ways to project future precipitation patterns. However, the spatio-temporal resolution for these GCMs is usually very large: Most of the information is available at the daily time scale and 1 degree longitude and latitude (about several hundreds of kilometers) spatial scale, with cutting-edge GCMs towards sub-daily and 30 km spatio-temporal scales [18]. These GCM scales are much larger than the scales of urban catchments. Engineering remediation and design typically consider urban sewersheds or subcatchments less than 10 square kilometers. For example, the 960 sq km Chicago metropolitan area comprises about 218 sewersheds or subcatchments ranging from 0.05 to 23 sq km, with a median of 3 sq km [19,20]. Small catchments translate to a small time of concentration [20,21], which in turn requires means to estimate short duration, small time step (from several minutes to several hours) rainfall with both accurate volume and within-storm variability [22]. Therefore, downscaling methods are needed to communicate the information from GCMs to the requirements of civil engineering study & design.

Dynamical downscaling, a method that nests Regional Climate Models (RCMs) within GCMs to simulate regional climate behaviors, has become a widely accepted way to downscale GCM outputs for small scale climate change impact studies. To get good estimates of small scale precipitation intensity that can be used for urban hydrological and hydraulics (H & H) studies, it is important for RCM to run at small spatio-temporal resolutions, especially for precipitation extremes [15,23,24]. Gutowski et al. [25] pointed out that 15 km spatial resolution resulted in 6-h precipitation intensities that attain some acceptable precision in the RCM. Mishra et al. [16] suggested that resolution in the order of of 2–5 km might help in better capturing the statistics of 3-h precipitation extremes. However, existing RCM simulations with these small levels of spatio-temporal resolutions are specialized to either a short time domain or small spatial domain, due to high computational costs [8,26].

Large discrepancies may occur in the behavior of sub-daily extreme precipitation when the required resolutions are not met. Over North America, this generally results in underestimation of precipitation at coarser resolution [16,27,28]. The magnitude of underestimation increases as the rainfall duration decrease [16,27,28]. Mishra et al. [16] analyzed 3 and 24-h extreme precipitations at urban areas of the United States using the 50-km resolution North American Regional Climate Change Assessment Program (NARCCAP) [29] RCMs. Ninety-six of the 100 areas they studied have NARCCAP simulated 3-h 100-year return period precipitation underestimating the observations. Only a few of the locations were found to have RCM-simulated 3-h and 24-h 100-year return period extreme precipitation within the ±10% error band of the observed estimates.

In addition to the biases, small scale precipitation projections from dynamical downscaling may also be associated with large uncertainties. The driving GCM outputs are carrying uncertainties from emission scenario, GCM model formulation, and natural variability [27,30,31]. These uncertainties are amplified when added with uncertainties of RCM and RCM-GCM pairing in the downscaling process [31]. With better representation of changes in the small-scale processes, reducing the spatio-temporal resolution of RCM simulations can reduce the bias and also decrease the uncertainties in projected extreme precipitation [27]. Many studies have emphasized the importance of natural variability [32,33] and variations in RCM formulation [34] to the uncertainties of projected small-scale precipitation extremes. In recent years, extensive cross-agency efforts have been put into multi-model ensemble systems that compare different RCMs and realizations, like the NARCCAP program [29] and the Coordinated Regional Climate Downscaling Experiment (CORDEX) [35]. Multi-model ensemble systems with long-term simulations from different GCM-RCM should be able to account for uncertainties from natural variability, GCM and RCM model formulations, and GCM/RCM pairing. They are one of the most reliable data sources for future climate projections and are widely used in climate change impact studies [16,27].

Many previous studies have looked at accessing small-scale future precipitation extremes and the uncertainties associated with them. However, we need a better understanding of how different models project small-scale extreme precipitation at different frequencies and durations, as well as the uncertainties associated with these projections at different frequencies and durations. The first part of this study aims at answering these two questions. Simulated 3-h rainfall time series from NARCCAP were used in frequency analysis to estimate rainfall quantiles from 2-year to 100-year return period, at durations of 3-h, 24-h, and 30-day. The Calumet Drop Shaft-51 (CDS-51) catchment in the Chicago Metropolitan area was selected as a case study. Biases and uncertainties were analyzed at different frequencies and durations, from simulations of different NARCCAP GCM/RCM combinations.

Even though there are new data sources that use more updated emission scenarios, or smaller spatial steps that can better capture the behavior of small-scale extreme precipitation, those data sources are limited to selected model, time domain, and locations. NARCCAP is the dynamically downscaled multi-model ensemble for North America that has the smallest time step (3 h) and spatial step (50 km). However, there is still a gap between the spatio-temporal scales required for urban H & H studies, and the ones of available multi-model ensembles. When looking at short-duration extreme precipitation from NARCCAP, Mishra et al. [16] found that 3-h precipitation maxima means and 100-year return period magnitudes at a large number of studied urban areas across the United States are "largely underestimated" for both reanalysis and GCM boundary conditions, with biases over 10% of the observed estimates. These biases are not appropriate to use directly for stormwater infrastructure design purposes. They ascribed this bias to the weakness of RCMs in the convective parameterizations and its spatial scale.

One widespread interpretation for future climate change among both climatologists and end users is that "By the end of the century, an Illinois summer may well feel like one in East Texas today" [36]. An underlying assumption of this saying is that, the change of climate in time, from now to future, can be represented by the change of climate in space, from one place to another. Inspired by this idea, this paper examines an alternative method called extreme precipitation spatial analog to represent potential future extreme precipitation at small spatio-temporal scale. The spatial analog method uses real climate records from other locations to represent potential future climate. This is a novel alternative to the climate model projections and provides an intuitive way to communicate potential future climate projections to climate change adaptation and mitigation studies [29]. "A spatial analog is a location whose climate during a historical period is similar to the anticipated future climate at a reference location" [37]. As a "local, concrete, and immediate" [38] way to access future climate, the spatial analog method is particularly useful for urban studies, most of which focus on using multiple climate variables to study the social or economic vulnerability of cities [39–42]. Previous studies that looked at precipitation spatial analogs have focused on average precipitation patterns at large time scales. Hallegatte et al. [40], Grenier et al. [37], and Kellett et al. [42] used annual precipitation as the metric to find spatial analogs. Kopf et al. [41] include 30-year Aridity Index as a metric. Hallegatte et al. [40],

Horváth et al. [43] and Ramírez-Villegas et al. [44] adopted monthly precipitation in their studies. Among rare attempts that studied precipitation spatial analog, Kellett et al. [42] looked at small temporal scale extreme precipitation of spatial analogs: Extreme daily precipitation of the whole study time periods were used as a metric for selecting spatial analogs. However, no previous study has focused only on looking at spatial analog for extreme precipitation, nor do they examine frequencies of sub-daily extreme precipitation in spatial analog. Extreme precipitation at sub-daily and shorter time scales have different mechanism to those at longer time scales [45]. The high variability of extreme precipitation at small spatio-temporal scales is crucial for urban hydrologic and hydraulic design. Existing studies mentioned the need for denser rain gauge network and finer resolution radar data [46,47]. However, due to their coarse resolution, current climate model ensemble projections are limited in their ability to represent small scale physics necessary to generate hydrologic data for urban H & H studies. The novelty of this approach is that it bridges the gap between projections from current climate model ensemble and fine spatial and temporal resolutions needed for H & H design by adopting the concept of spatial analog to extreme precipitation. When we borrowed the concept of spatial analog to focus only on extreme precipitation, the underlying hypothesis is that the future extreme storm characteristics of a certain location will be similar to the current extreme storm characteristics of a different location. Since extreme precipitation at different durations and frequencies are of interest for various urban H & H studies depending on their spatio-temporal scales, it is of great importance to study extreme precipitation spatial analog at different durations and frequencies.

In this study, we examine if an extreme precipitation spatial analog method is feasible to project future precipitation extremes at the scales required for urban H & H studies. Here we use the intensity-duration-frequency (IDF) relation as a metric for extreme rainfall spatial analog. The CDS-51 catchment in the Chicago Metropolitan area was selected as a case study to test the method. Given that current large model ensembles of climate model simulated dynamically downscaled rainfall have not yet reached the hourly and sub-hourly resolution, evaluating the possibility of using other cities' high-resolution rainfall data to represent future precipitation of CDS-51 can be valuable for further climate change impact studies on urban H & H processes. Since CDS-51 is located in the Chicago metropolitan area of Illinois, this study can also evaluate the original concept of the transformation from temporal change to spatial change from Kling et al. [36], who used Illinois as an example to illustrate the migrating of climate.

Previous studies have showed that the intensity and duration of small duration heavy precipitation in the Chicago area are projected to increase [28,48,49]. Markus et al. [28] used rainfall outputs from one GCM and RCM combination with 30 km resolution to analyze the rainfall frequency quantiles for 2046–2055. Results showed that 24-h extreme precipitations of 2-year, 5-year, and 10-year return periods exhibited an average 20% and 16% increase under the Special Report on Emissions Scenarios (SRES) A1FI and B1 scenario [50] in the northern part, and a minor increase of 3% under A1FI scenario and a decrease of 12% under B1 scenario for the southeastern part. These results were updated by the same research group in 2018 [51], with projected increase in 24-h rainfall of 2-year to 100-year return period under SRES A1B, A2, and B1 scenarios using statistically downscaled 13 Coupled Model Intercomparison Project Phase 3 (CMIP3) models at 10 km resolution [52].

Dynamically downscaled model ensembles for North America using Coupled Model Intercomparison Project Phase 5 (CMIP5) data [53], like the North American Coordinated Regional Downscaling Experiment (NA-CORDEX) [54], do not have enough simulations from different RCM/GCM combinations at temporal scale as small as 3-h yet. Wuebbles et al. [55] have found that there are large variances between simulated extreme precipitation of different models in CMIP5 data. Extreme precipitation spatial analog derived using precipitation projections from only a few of CMIP5 models would be very uncertain. In addition, Wuebbles et al. [55] showed that the CMIP5 overall projected changes of extreme precipitation are in general similar to those of CMIP3 [56]. Specifically, Markus et al. [51] compared dynamically and statistically downscaling and also CMIP3 and CMIP5 projected daily extreme precipitation at the Chicago area and did not find significant difference. Taking all these factors into consideration, NARCCAP [29], which provide 3-h 50-km resolution rainfall simulations from over 10 combinations of RCM/GCM with SRES A2 scenario [50], is one of the best data sources for future extreme rainfall studies at the regional scale, and was used in this study to test the possibility of extreme precipitation spatial analog method. Future study can be conducted with more advanced multi-model ensembles to find extreme precipitation spatial analog using the method provided in this study.

2. Data and Methods

A schematic plot for the methodology of this study is shown in Figure 1. First we obtained the NARCCAP simulated rainfall data at the study area using four different remapping methods (see Sections 2.1–2.3). Then we calculated the annual maximum series (AMS) of NARCCAP rainfall at 1968–2000 and 2041–2070. Regional frequency analyses were conducted using these AMS to get frequency quantiles of NARCCAP rainfall data (see Section 2.4). We also sampled all the National Oceanic and Atmospheric Administration (NOAA) Atlas 14 (NA14) rainfall quantiles of the study area. After that, these NARCCAP rainfall frequency quantiles were bias-corrected with the ratio method using NA14 quantiles of the study area (see Section 2.5). Bias-corrected future NARCCAP rainfall quantiles were then compared to NA14 relations at other cites in search of potential spatial analog.

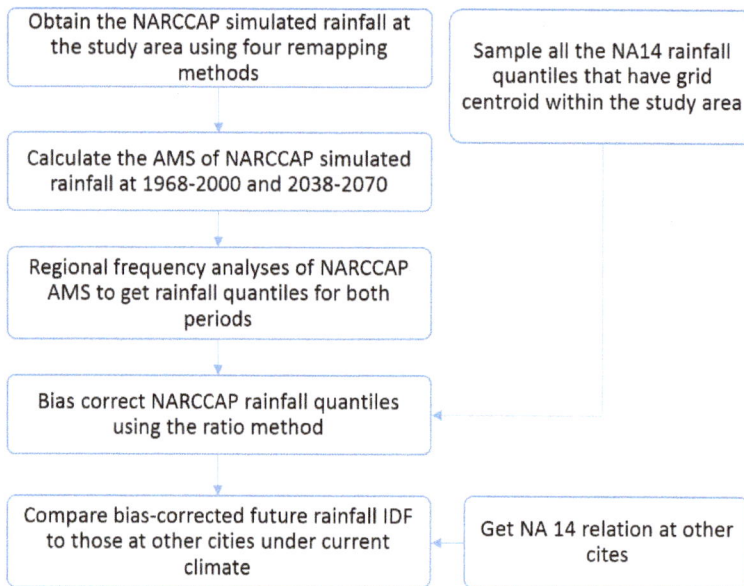

Figure 1. Schematic of the methodology in this study.

2.1. Study Site

The CDS-51 catchment (see Figure 2), which belongs to the Calumet system of Chicago's Tunnel and Reservoir Plan (TARP) [57], has been widely used in H & H studies [20,58,59]. The CDS-51 catchment is a 5th-order complex urban system with an area of 3.16 km^2 in the Village of Dolton, IL. The spatial characterization of the CDS-51 catchment was collected and analyzed during the Phase II of TARP project [20]. The critical duration for this catchment is 1 h [19].

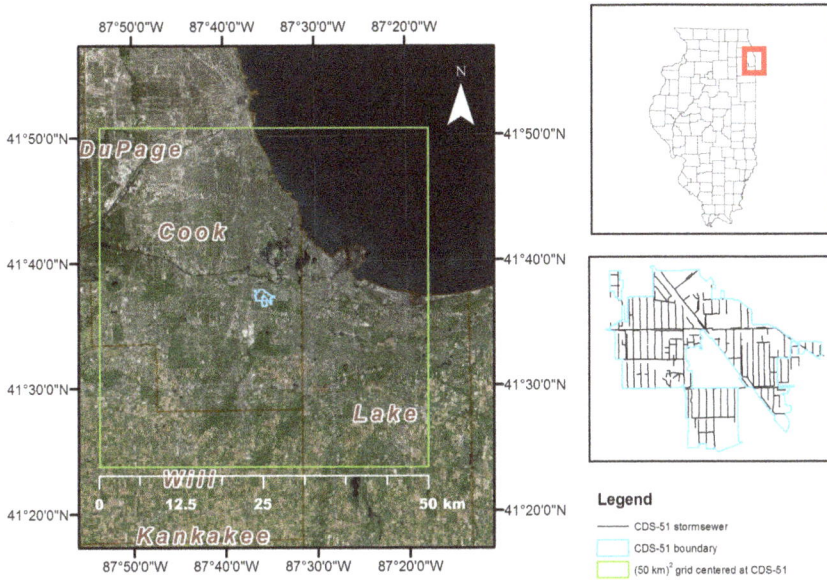

Figure 2. Location of the study area, with the boundary of the study grid (the green line) and the boundary of CDS-51 catchment (the blue line).

2.2. Precipitation Data

The climate model simulated rainfall used in this study come from NARCCAP [29]. NARCCAP is an international program that aims at producing high resolution climate change simulations for investigating uncertainties in future climate projections and climate change study at regional scale for the North American region [29,60]. It provides dynamically downscaled precipitation at small spatial-temporal resolution of 50-km and 3-h, from different combinations of GCM and RCM. The 10 RCM/GCM combinations used in this study are shown in Table 1. The four GCM simulations were obtained from the CMIP3 archive [56]. The driving emission scenario for the GCM simulations is the SRES A2 scenario [50]. Under Phase II of NARCCAP, the five RCMs were run for the current period of 1968–2000 and future period of 2041–2070, with boundary conditions provided by those GCMs. Despite the fact that there are new GCM model simulations under updated climate models and emission scenarios, like the CMIP5 models and Representative Concentration Pathways (RCP) scenarios [18], in this study we chose the NARCCAP dataset with CMIP3 GCMs. As we argued previously, small spatial and temporal resolutions are needed for climate models to simulate the small-scale processes that are associated with urban catchment scale precipitation extremes. At the same time, to account for uncertainties and biases from model formulation and natural variability, multi-model ensemble is preferred to single model runs. Unfortunately, due to the large computation expense, dynamically downscaled small resolution multi-model ensembles using models from the CMIP5 archive under RCP emission scenarios are still not available for the entire United States. NARCCAP is the best dataset publicly available, dynamically downscaled to sub-daily time scale, verified and applied by many studies, and with a number of RCM/GCM combinations.

Table 1. RCM/GCM combinations from NARCCAP [29] used in this study.

No.	RCM	GCM
1	CRCM	ccsm
2	CRCM	cgcm3
3	HRM3	gfdl
4	HRM3	hadcm3
5	MM5I	ccsm
6	MM5I	hadcm3
7	RCM3	cgcm3
8	RCM3	gfdl
9	WRFG	ccsm
10	WRFG	cgcm3

To evaluate the climate models' performance, we compared the model simulated rainfall's intensity-duration-frequency (IDF) relation under present-day climate with those of observations. The most recent nationwide calculation of rainfall quantiles, National Oceanic and Atmospheric Administration (NOAA) Atlas 14 (NA14) [61], was adopted in this study to represent the observed climate conditions. The NA14 rainfall data time periods are 1948–2000 for hourly rainfall measurements, and 1863–2000 for daily rainfall measurements. We looked at 3-h, 24-h, and 30-day precipitations from NA14. Rainfall quantiles with duration under 3-h were not used in this study, since NARCCAP cannot be used to provide rainfall quantiles at equivalent durations. Return periods examined are 2, 5, 10, 25, 50 and 100 years. IDF relations at the location of the study site and four cites south and west of Chicago (where the study site is located) were also obtained. The four cites are: Champaign IL, Springfield IL, St. Louis MO, and Memphis TN (see Figure 3). As can be seen in Figure 3, all five cities fall within 2.5 degree longitude of each other. Hence, we anticipate them to have similar climate. As locations shift south from Chicago to Memphis, we expected to see increases in precipitation extremes that could portray the changes in time projected for Chicago.

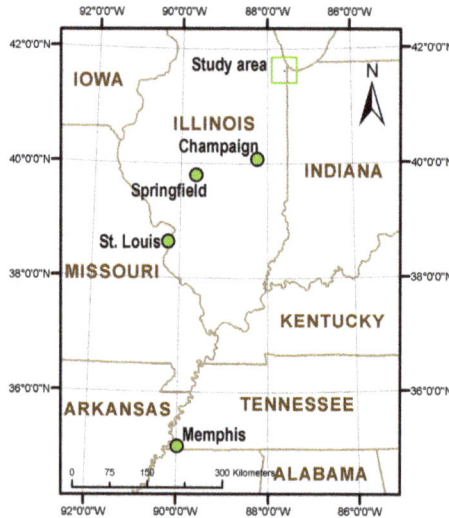

Figure 3. Location map of the study area, Champaign IL, Springfield IL, St. Louis MO, and Memphis TN.

2.3. Regridding and Sampling of Rainfall Data

To account for the differences between spatial resolutions of NA14 precipitation dataset and NARCCAP simulated precipitation, the study area of this paper was set to be a rectangular grid centered at the centroid of CDS-51 catchment and has the same resolution of the NARCCAP horizontal grids (i.e., 50 km), which is shown as the area with the green boundary lines in Figure 2. This area is referred to as "the study area" or "the study grid".

Different RCM/GCM combinations of NARCCAP have different horizontal grids systems even though they are in the same resolution. As a result, regridding or remapping is needed to interpolate horizontal fields of all model combinations to our study grid. Some research indicates that regridding might cause smoothing in the extremes [62,63], but it changes with a variety of factors including interpolation or averaging approach, number of points, etc. In this study, we quantified the potential uncertainties in regridding process by using four widely-recognized remapping methods: bilinear interpolation, bicubic interpolation, distance-weighted average remapping, and nearest neighbor remapping. The Climate Data Operators (CDO) [64] was used for the regridding process. Although we recognize the potential smoothing of extremes from regridding, we anticipate that the impact of smoothing is smaller than that of difference among different NARCCAP model grids, and therefore regridded the data to a consistent study grid. We include nearest-neighbor approach as this does not smooth the extremes.

As the study area is much larger than NA14 horizontal grids (0.0083 degrees longitude and latitude or around 900 m) [61], all NA14 grids that have their centroids inside the study area were obtained to provide the rainfall quantiles under present-day climate.

2.4. Precipitation Frequency Analysis and IDF Relation

Precipitation frequency analyses were conducted for NARCCAP simulations under historical and future periods using the L-moments method to get the IDF relations of rainfall. By picking the maximum rainfall depth of a certain duration for each year, AMS were formed for each of the selected durations. These AMS were then fitted into distributions using the L-moments method. The L-moments method [65] is widely used in precipitation frequency analyses, including the NA14 [61]. For this study, the L-moments are used instead of conventional moments, due to several reasons. First, L-moments "characterize a wider range of distributions"; second, when estimated from a sample, they are "more robust to the presence of outliers in the data"; third, L-moments are "less subject to bias in estimation" [65]. Details about the L-moments and frequency analysis using L-moments can be found in Hosking and Wallis [65]. In their book, Hosking and Wallis [65] defined L-moments λ_r as linear combinations of Probability Weighted Moments (PWM) β_r. Greenwood et al. [66] defined the rth order PWM of a random variable X with cumulative distribution fuction $F(X)$ to be $\beta_r = E[X\{F(X)\}^r], r = 0, 1, 2, \ldots$. An unbiased estimator b_r of β_r given by Hosking and Wallis [65] is $b_r = n^{-1}\binom{n-1}{r}^{-1} \sum_{j=r+1}^{n} \binom{j-1}{r} x_{j:n}$, where $x_{j:n}$ denotes the jth smallest value in a sample of size n. The relationship between PWM β_r and popular L-moments λ_r are: L-location $\lambda_1 = \beta_0$, L-scale $\lambda_2 = 2\beta_1 - \beta_0$, $\lambda_3 = 6\beta_2 - 6\beta_1 + \beta_0$, $\lambda_4 = 20\beta_3 - 30\beta_2 + 12\beta_1 - \beta_0$. Popular L-moment ratios τ_r can be calculated by: L-CV $\tau = \lambda_2/\lambda_1$, L-skewness $\tau_3 = \lambda_3/\lambda_2$, and L-kurtosis $\tau_4 = \lambda_4/\lambda_2$. The R package lmom [67] was used to conduct this analysis.

The purpose of this research is to explore the spatial analog method in extreme precipitation. The type of distribution that provides the best fit for extreme precipitation may be different among different models and durations, consequenctly the best fit frequency curve for each NARCCAP model precipitation series for each duration were selected from among five types of distributions: Generalized Logistic (GLO), Generalized Extreme Values (GEV), Generalized Normal (GNO), Generalized Pareto (GPA), and Pearson Type III (PE3) (see Table 2). Some studies recommend the GEV above other distributions for extreme precipitation frequency analysis [61,68,69]. But our comparison showed that the difference between two methods' resulting quantiles are very small (on average −0.4%).

The distributions with the smallest Root Mean Square Error (RMSE) were selected as the best fit distributions in this study [70–72]. Rainfall quantiles with the six different return periods selected for this study were then calculated using the quantile functions of the chosen distribution for each duration and RCM/GCM model combination.

Table 2. Summary of equations to calculate the characterizing parameters: ζ (location), α (scale), and k (shape) with population L-moments: L-location λ_1, L-scale λ_2, L-CV τ, L-skewness τ_3, and L-kurtosis τ_4 [65].

Dist.	ζ	α	k
GLO	$\zeta = \lambda_1 - \alpha\left(\frac{1}{k} - \frac{\pi}{\sin k\pi}\right)$	$\alpha = \frac{\lambda_2 \sin k\pi}{k\pi}$	$k = \tau_3$
GEV	$\zeta = \frac{\lambda_1 - \alpha[1-\Gamma(1+k)]}{k}$	$\alpha = \frac{\lambda_2 k}{(1-2^{-k})\Gamma(1+k)}$	$k = 7.8590 + 2.9554c^2,$ $c = \frac{2}{3+\tau_3} - \frac{\log 2}{\log 3}$
GNO	Numerical methods were used to estimate GNO distribution using L-moments [65].		
GPA	$\zeta = \lambda_1 - (2+k)\lambda_2$	$\alpha = (1+k)(2+k)\lambda_2$	$k = (1-3\tau_3)/(1+\tau_3)$
PE3	$\zeta = \lambda_1$	$\alpha = \frac{\lambda_2 \pi^{1/2}\sigma^{1/2}\Gamma(\sigma)}{\Gamma(\sigma+\frac{1}{2})}$	$\sigma = \frac{0.36067z - 0.59567z^2 + 0.25361z^3}{1-2.78861z+2.56096z^2-0.77045z^3},$ $z = 1 - \|\tau_3\|, k = 2\sigma^{-1/2}\sin(\lambda_3)$

Previous studies have mainly used two methods to extract extreme precipitation data for frequency analysis: Annual Maximum Series (AMS) and Partial Duration Series (PDS) [51,61]. AMS are formed by extracting the maximum values from each year of record. The PDS method, on the other hand, allows more than one extreme value to be extracted from a single year. The widely-adopted Chow et al. [73] method [51,61] (Equation (1)) was used to convert the T_{PDS} return periods from NA14 data to the T_{AMS} from the model projections. In this study, the T_{PDS} we analyzed are 2, 5, 10, 25, 50, and 100 years. The corresponding T_{AMS} were calculated using Equation (1), which were used later in frequency analysis of AMS to generate unbiased quantiles at our T_{PDS} of interest. For example, $T_{AMS} = 2.54$ when $T_{PDS} = 2$ using Equation (1). This means that a 2-year unbiased quantile can be estimated by a 2.54-year AMS-based quantile.

$$T_{AMS} = \frac{1}{1 - e^{-\frac{1}{T_{PDS}}}} \tag{1}$$

2.5. Bias Correction

As discussed previously, climate model simulated precipitation can be biased. Correcting for biases in climate model output using historical data for future projections assumes that the past performance of climate model could be an indicator of future skill. Even though there are still debates on this issue, many studies have proved the existence of systematic errors in climate model outputs and showed some evidence supporting this assumption [51,74–76]. Bias correction was applied to NARCCAP future quantiles. The bias correction method used in the paper is the ratio method [28], also referred as the multiplicative change factor method [77]. The ratio method is a widely used linear correction method for extreme precipitation quantiles [28,78]. To get bias-corrected future quantile, P_f^*, the ratio method multiplies the ratio between NA14 quantile O and model simulated quantiles of current climate P, to the model simulated quantiles of future period P_f. In this study, this correction was conducted for every NARCCAP quantiles at each selected return period and duration, using all NA14 quantiles within the study area, as shown in Equation (2). All NARCCAP historical quantiles from different RCM, GCM, and grid-remapping methods, at different durations and return periods, were bias-corrected separately. Each of these NARCCAP historical quantiles were bias-corrected to all NA14 quantiles within the study grid. In this way, the spatial variations of NA14 quantiles at the scale of our study area are taken into the consideration by the bias-corrected future NARCCAP quantiles.

$$P_{f,ijkl}^* = \alpha_{ijkl} * P_{f,ijl} \tag{2}$$

in which $\alpha_{ijkl} = O_{ijk}/P_{ijl}$, is the bias-correction ratio, i stands for different durations, j stands for different return periods, k stands for the different NA14 grids within the study area, l stands for the different combinations of RCM, GCM, and remapping methods for NARCCAP quantiles.

3. Results

The results for frequency analysis are based on a grid centered at the CDS-51 catchment with the same horizontal resolution of a NARCCAP grid (50 km×50 km). Rainfall frequency quantiles from NARCCAP simulations and NA14 were analyzed at rainfall durations of 3-h, 24-h, and 30-day.

3.1. Rainfall Frequency under Current Climate

Figure 4 shows the rainfall frequency quantiles from NA14 of the study grid. The range of NA14 frequency quantile estimates within the study grid (the grey shaded area) reveals the spatial variance of rainfall statistics at the spatial resolution of NARCCAP data at CDS-51. At the 3-h rainfall duration, this spatial variance is much smaller than those at 24-h and 30-day durations. The spatial variance is also increasing with decreasing exceedance probability (or the increase of return period). This trend is more obvious in 24-h and 30-day rainfall than in 3-h rainfall. At 100-year return period, the spatial variance of NA14 quantiles can be as high as 20 mm for 24-h and over 30 mm for 30-day rainfall. In addition to the variance caused by the spatial scale of the study grid, the confidence intervals of the frequency quantiles also are increasing slightly with decreasing of exceedance probability.

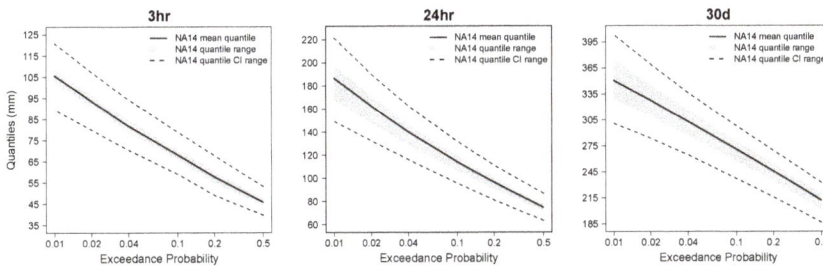

Figure 4. Frequency quantiles from observation data for the study area, for 3-h, 24-h, and 30-day rainfalls.

3.2. Modeled Historical Period Rainfall Quantiles vs. Current Rainfall Quantiles

Figure 5 shows the comparison between frequency quantiles from NARCCAP simulations of historical periods (1968–2000) and NA14 for the study grid. There are 40 NARCCAP realizations in total, from 10 RCM/GCM combinations, each regridded using four remapping methods. The uncertainties that come from GCM, RCM, and grid remapping for the NARCCAP frequency quantiles are extremely large compared to the range of NA14 frequency quantiles within the study grid. These uncertainties increase at longer return periods. At the 3-h duration, most of the NARCCAP historical frequency quantiles underestimate the observed NA14 frequency quantiles, except for those with return periods larger than 50-year. For 24-h rainfall, the NA14 frequency quantiles generally lie at the upper bound of the NARCCAP historical quantiles range. Our results are comparable to the findings of Mishra et al. [16], that NARCCAP RCM underestimate 3-h and 24-h extreme precipitations in the east part of the United States. However, for the 30-day duration events, the NA14 quantiles are within the variability of the NARCCAP historical quantiles. The large uncertainty in the model-simulated rainfall quantiles is the motivation for using bias-correction. As will be discussed below, using bias-correction we are able to easily quantify changes to the different quantiles, despite the biases that may exist for individual model simulations.

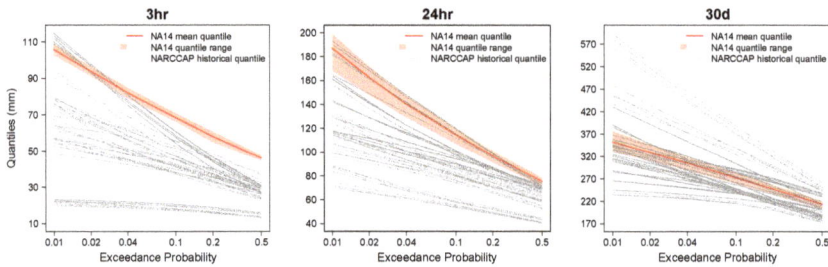

Figure 5. Frequency quantiles from NARCCAP historical period (1968–2000) simulations and NA14, of the study grid, for 3-h, 24-h, and 30-day rainfalls.

3.3. Bias-Correction of Modeled Frequency Quantiles

Bias-correction forces the frequency quantiles of NARCCAP historical time period to match the NA14 quantiles. By multiplying the ratio of each NA14 quantile within the study grid to each of the NARCCAP quantiles, all NARCCAP quantiles from different RCM/GCM combinations, remapping methods, at different return periods and durations, coincided with the specific NA14 quantile used for bias-correction. As shown as an example for 3-h rainfall in Figure 6, the mean of all bias-corrected NARCCAP historical quantiles coincide with the mean of NA14 quantiles at all exceedance probabilities. The mean bias-correction ratios for each GCM and RCM combination are shown in Figure 7. These ratios can serve as evaluations of the performance of RCM and GCM. Values larger than 1 indicate that the particular NARCCAP model is underestimating precipitation, while values less than 1 indicate overestimation. Some models perform better at smaller durations, while others better at longer durations. For example, HRM3 model frequency quantiles are the closest ones to NA14 quantiles at 3-h and 24-h durations, but for 30-day duration they tend to overestimate the NA14 quantiles. CCSM model results largely underestimate at 3-h and 24-h durations, but their results at 30-day duration are realistic. Models can also perform differently at different return periods. In addition, the correction ratios tend to depend more on the RCM than the GCMs: Cells with the same RCM have more similar colors than cells with the same GCM (Figure 7). This result matches with the finding of Monette et al. [79] and Khaliq et al. [80] who studied seasonal 1-day to 10-day extreme precipitations for Canadian watersheds using NARCCAP data. The control of RCM on the performance seems to gets weaker at 30-day duration, where ratios of the same RCM model can be very different from each other. Taking HRM3 again as an example, at 30-day duration, ratios of HRM3+gfdl have a relatively good performance, with more overestimation at larger exceedance probabilities, while ratios of HRM3+hadcm3 are giving the worst results among all the model combinations, with more overestimation at smaller exceedance probabilities. This overestimation of HRM3 was also observed by previous study of Monette et al. [79]. They found that HRM3 tends to overestimate multiday rainfall extremes, with more bias at longer duration and return period.

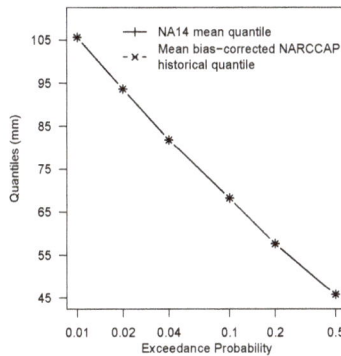

Figure 6. Mean frequency quantiles from bias-corrected North American Regional Climate Change Assessment Program (NARCCAP) historical period (1968–2000) simulations and Atlas 14 (NA14), of the study grid, for 3-h rainfalls.

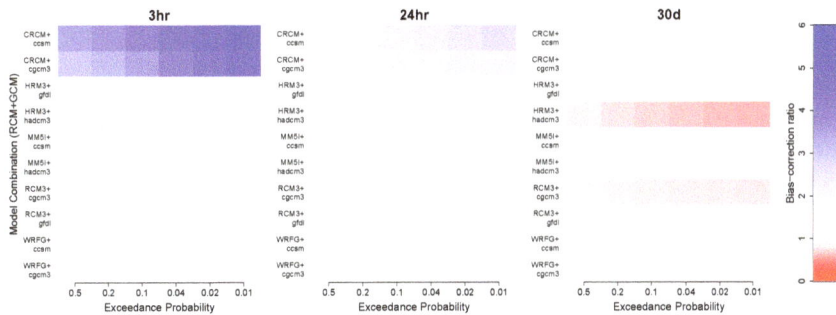

Figure 7. Bias-correction ratios for NARCCAP frequency quantiles (NA14/NARCCAP) averaged for each General Circulation Model (GCM) and Regional Climate Model (RCM) combination, for 3-h, 24-h, and 30-day rainfall, at exceedance probabilities from 0.5 to 0.01 (2-year to 100-year return period).

3.4. Bias-Corrected Modeled Future Frequency Quantiles

Figure 8 shows the average behavior of future bias-corrected NARCCAP frequency quantiles and NA14 frequency quantiles of the study area. There is a clear increase in rainfall depth in the bias-corrected future mean NARCCAP quantiles for all three durations from 3-h to 30-day, when compared to the NA14 quantiles. For 3-h rainfall, this increase of rainfall depth is almost an identical increase of about 8 mm at different exceedance probabilities from 0.01 to 0.5. At the 24-h duration, the difference between mean bias-corrected future quantiles and NA14 quantiles decreases for longer return periods. On the other hand, at the 30-day duration, this difference increases slightly at longer return periods, with a maximum difference of over 20 mm. Even though these differences between "current" and mean projected future frequency quantiles are obvious, they may not be statistical significant: Most of the bias-corrected NARCCAP future quantiles are still within the confidence intervals of the NA14 quantile range. The only exception is at 3-h duration and exceedance probability larger than 0.2. In general, at larger exceedance probabilities the mean bias-corrected future quantiles are closer to the upper boundary of confidence limits of the "current" frequency quantiles range of the study area.

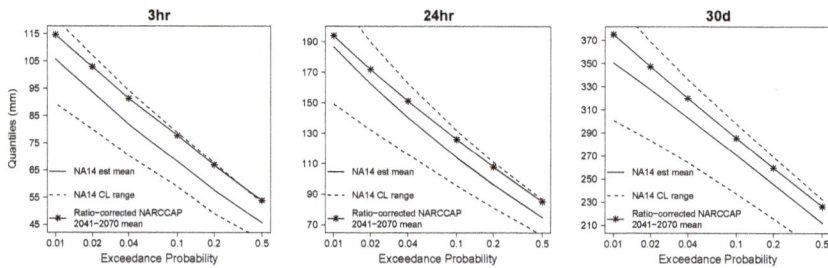

Figure 8. Frequency quantiles from NA14 and bias-corrected NARCCAP 2041–2070 period of the study area, for 3-h, 24-h, and 30-day durations.

A Welch Two Sample t-test analysis on the difference of the means and a Wilcoxon Rank-Sum test of the two-sample medians indicated that the means and medians of bias-corrected NARCCAP future (2041–2070) quantiles are different from NA14 quantiles of the study area at all analyzed exceedance probabilities and durations. Welch Two Sample t-test is an unpaired two-sample location test with a null hypothesis that the two populations have equal means. It assumes normality and is more reliable than Student t-test when the two samples have unequal variances. Wilcoxson Rank-Sum test is a non-parametric test with a null hypothesis that the two populations have equal median. It does not assume normality but does assume equal variance. By using both tests, we can perform robust statistical inference on the difference between average behaviors of the bias-corrected NARCCAP future quantiles and NA14 quantiles. The results from both tests indicate that the average projected future rainfall quantiles are larger than those under current climate for the study area with high confidence.

Figure 9 shows how the range of future bias-corrected NARCCAP frequency quantiles and NA14 frequency quantiles of the study area compare to each other. The range of bias-corrected NARCCAP future quantiles are always much larger than the NA14 quantiles. Furthermore, the NA14 ranges are within the NARCCAP ensemble for all three durations. The NA14 range increases with longer return periods and longer rainfall durations from 3-h to 30-day. Bias-corrected future NARCCAP ranges also increase for longer return periods and rainfall durations. Figure 9 also shows all the realizations of future bias-corrected NARCCAP quantiles compared to NA14 quantiles of the study area. The darkness of the grey color represents the density of future NARCCAP realizations. In general, NARCCAP bias-corrected future quantiles are higher than NA14 at all frequencies, especially at smaller durations of 3-h and 24-h. At the duration of 30-day, with the extreme high values of MM5I model at longer return periods, the mean future NARCCAP quantiles are higher than NA14 quantiles (as shown in Figure 8).

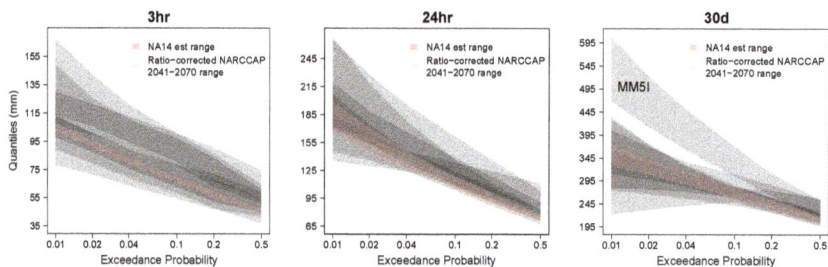

Figure 9. All frequency quantiles' realizations from NA14 and bias-corrected NARCCAP 2041–2070 period of the study area, for 3-h, 24-h, and 30-day durations.

3.5. Rainfall Frequency Quantiles from Other Cities under Current Climate

NA14 quantiles from several other cities south of Chicago were used to compare with the projected future changes in rainfall. Figure 10 shows how bias-corrected NARCCAP future frequency quantiles of the study area compare to the NA14 quantiles of those other cities south of the study area. Figure 11 shows the differences between the quantiles from NA14 and mean bias-corrected NARCCAP future quantiles. For 3-h rainfall, the NA14 quantiles from Memphis is the closest one to the mean bias-corrected NARCCAP future quantiles. At 24-h duration, NA14 quantiles from St. Louis are the nearest to NARCCAP bias-corrected future mean quantiles of the study area. For 30-day duration, NA14 quantiles from Springfield and Champaign are closest to NARCCAP bias-corrected future mean quantiles of the study area.

The result supports our hypothesis that the change of extreme precipitation in time, from now to future, can be approximated by the change of extreme precipitation in space, from one location to others. This implies that the spatial analog method is feasible to utilize observed precipitation data from one location to represent potential future extreme precipitation. For example, Memphis can be used as 3-h extreme precipitation spatial analog for the study area; St. Louis can be used as 24-h extreme precipitation spatial analog for the study area; Springfield and Champaign can be used as 30-day extreme precipitation spatial analog for the study area. Spatial analog of extreme precipitation will likely depend on the duration of rainfall of interest.

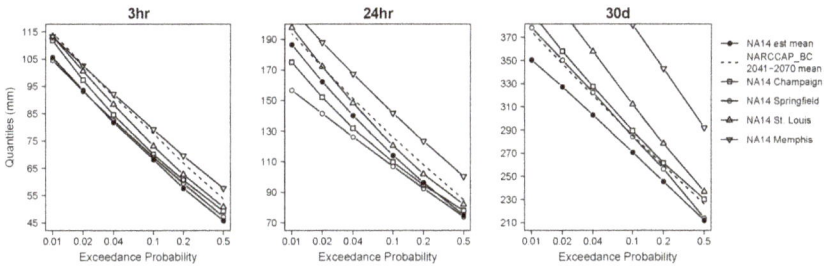

Figure 10. Mean frequency quantiles from NA14 and bias-corrected NARCCAP 2041–2070 period of the study area compare to NA14 quantiles from other cities south of the study area, for 3-h, 24-h, and 30-day durations.

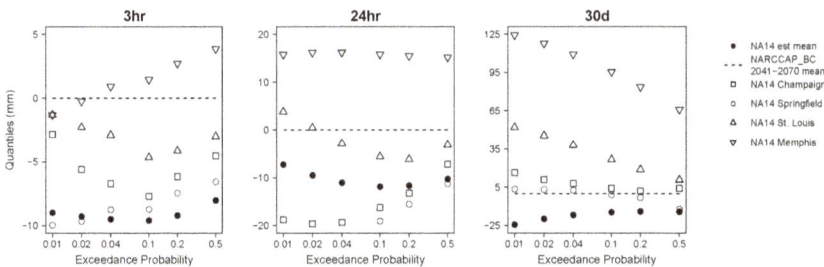

Figure 11. Differences between mean NA14 frequency quantiles of the study area as well as the other cites, and mean bias-corrected frequency quantiles from NARCCAP 2041–2070 period of the study area, for 3-h, 24-h, and 30-day durations.

4. Conclusions and Discussion

It has been hypothesized that as climate warms, the temperature and precipitation in a location such as Illinois could be more like the climate in a more southern location like Texas [36]. This implies that we could potentially use spatial analogs, instead of relying on GCM projections of future climate. This is particularly attractive in the case of urban water infrastructure, because of the very small spatial and temporal scales required for the design of stormwater systems. We test the spatial analog hypothesis by analyzing the observed extreme precipitation, characterized using NA14 IDF relations, at 3-h, 24-h, and 30-day durations and 2-year to 100-year return periods, in a study grid that centered at Calumet Drop Shaft-51 catchment of the Chicago metropolitan area. We compare these observations with future projections of extreme precipitation as simulated using bias-corrected IDF relations from RCM-GCM combinations from the NARCCAP ensemble under SRES A2 scenario for the period 2041–2070. We then compare the NARCCAP projections with observed extreme precipitation in four other more southern cities: Champaign, Springfield, St. Louis, and Memphis. Although NARCCAP simulations are used in this study the analysis can be easily applied to updated RCMs as these data become more widely available.

The major findings of this study are:

1. The variances in raw NARCCAP historical frequency quantiles from different combinations of RCM, GCM, and grid remapping methods are much larger than the variance of NA14 quantiles within the study area for all three durations analyzed. Most NARCCAP historical realizations tend to underestimate NA14 quantiles at smaller durations (3-h and 24-h), while at 30-day NA14 quantiles lie within the NARCCAP quantile range. Consequently, bias-correction is required to correct for the NARCCAP biases.
2. The bias-correction ratio can serve as a metric for evaluating model performance. There is no individual model combination that best captures the behavior of NA14 rainfall quantiles at all durations and frequencies. The model performance depends on rainfall duration and return period. The correction ratios tend to depend more on the RCM than the GCM, especially at smaller rainfall durations of less than 24 h.
3. The projections of future extreme precipitation using bias-corrected NARCCAP 2041–2070 quantiles under A2 scenario have ensemble means and medians that are all statistically higher than the NA14 quantiles, indicating a future intensification of extreme events at all durations and return periods. For individual NARCCAP realizations, most of them are higher than NA14 at all quantiles. The uncertainty in NARCCAP bias-corrected future quantiles is still very large, compared to the variance of NA14 quantiles. This uncertainty increases with return period and rainfall duration.
4. We find that future 3-h rainfall in Chicago will be more similar to that of current-day Memphis, while longer-duration 30-day rainfall will be more similar to that of Springfield, IL. This indicates that the spatial analog is potentially a useful method to obtain future extreme precipitation projections, but highlights the fact that the analogs will likely depend on the duration of rainfall of interest.

Even though this study used a single case in the Chicago area, we speculate that the dependence of extreme precipitation spatial analog on the duration of rainfall found in this study is likely to appear in other cases (especially in Midwestern US) as well. The intensity of extreme storms in the Midwestern US has been increasing in the historical period [81], and these changes are dominated by a shift in the most intense precipitation, usually associated with mesoscale convective systems (MCS) that show an upward trend in intensity along the northern plains [82]. This means that regions in Iowa and Illinois are experiencing more intense short-duration events, that look more like the storms at locations further south (Memphis). However, on average, not all types of storms are increasing at this rate. Average precipitation includes stratiform precipitation, air mass convective precipitation, etc., and these types of storms do not seem to be increasing in intensity. So as we average in time, these trends in MCS storms will be "washed out" and our spatial analogs will be more like nearby locations (St. Louis or Champaign).

This study provides a first look at the possibility of extreme precipitation spatial analog to represent future precipitation projection at the scales of urban H & H study. Observed precipitation data from these spatial analogs that describe the behavior of extreme precipitation at time scales as small as sub-hourly, and spatial scale of 10 km or less, is a promising way to supplement the variability of localized extreme precipitation to the output of climate models. Other similar studies are needed to generalize our conclusions, and further investigate biases and limitations of the approach. In particular, more future studies are needed to explore how to adopt extreme precipitation spatial analog method in urban H & H study. One potential way is to use the real rainfall data with actual built-in time variability that satisfy the criteria of projected quantiles (i.e., total depth, duration, and frequency) as input to urban H & H models.

In addition, because the spatial analogs change with precipitation duration, further study is needed to examine whether the analogs identified for the shortest RCM duration analyzed (3 h in this study) are appropriate to extrapolate to the temporal scales required for urban catchments. As was mentioned earlier regarding duration and location of extreme precipitation spatial analog, we hypothesize that the longer durations reflect averaging among different types of storms. Although the temporal resolution of current generation models does not allow us to test this, we speculate that storms of shorter durations will tend to be associated with MCS and hence behave more similarly to each other than to longer duration storms. The next generation of models [54] should allow climate projections at shorter durations and once readily available could be used to test this hypothesis and identify the durations below which the analog tends to stabilize.

Nowadays, engineering practices are expected to design with consideration of the risk with its service life. There is a great need to design infrastructure economically not only to serve under current climate conditions, but also under potential future conditions. The uncertainty might be very large, but on the other hand, the benefit for any potential information on future precipitation is also very large. Even though there are large uncertainties associated with climate model projected extreme precipitation at spatio-temporal scales of urban studies, it is still probably the only method to project the future conditions at the temporal and spatial scales required for engineering design. As climate scientists are pushing hard on new generations of models, we hope that this bias and uncertainties can be gradually removed in the future.

Author Contributions: Conceptualization, all; methodology, all; formal analysis, A.K.W.; writing—original draft preparation, A.K.W.; writing—review and editing, all.

Funding: This research was funded by National Science Foundation (NSF) (award number 1331807) and United States Army Corps of Engineers(USACE)/United States Geological Survey (USGS) (award number G17AP0030).

Acknowledgments: We wish to thank the North American Regional Climate Change Assessment Program (NARCCAP) for providing the data used in this paper. NARCCAP is funded by the National Science Foundation (NSF), the U.S. Department of Energy (DoE), the National Oceanic and Atmospheric Administration (NOAA), and the U.S. Environmental Protection Agency Office of Research and Development (EPA). We also wish to thank NOAA/National Weather Service for providing data in NOAA Atlas 14 used in this paper. In addition, we would like to thank Joshua Cantone, Optimatics Vice President, for his hydrological studies on CDS-51 catchment, and Momcilo Markus, James Randal Angel, from the University of Illinois at Urbana-Champaign, as well as Shu Wu, from the University of Wisconsin-Madison, for help in their field of expertise in various stages of this research.

Conflicts of Interest: The authors declare no conflict of interest.

References

1. World Health Organization & UN-Habitat. *Global Report on Urban Health: Equitable Healthier Cities for Sustainable Development*; Technical Report; World Health Organization: Geneva, Switzerland, 2016.
2. United Nations, Department of Economic and Social Affairs, Population Division. *World Population Prospects: The 2017 Revision, Key Findings and Advance Tables*; ESA/P/WP/248; United Nations, Department of Economic and Social Affairs, Population Division: New York, NY, USA, 2017.

3.   Willems, P.; Olsson, J.; Arnbjerg-Nielsen, K.; Beecham, S.; Pathirana, A.; Gregersen, I.B.; Madsen, H.; Nguyen, V.T.V. *Impacts of Climate Change on Rainfall Extremes and Urban Drainage Systems*; IWA Publishing: London, UK, 2012.

4.   Winters, B.A.; Angel, J.R.; Ballerine, C.; Byard, J.L.; Flegel, A.; Gambill, D.; Jenkins, E.; McConkey, S.A.; Markus, M.; Bender, B.A.; et al. *Report for the Urban Flooding Awareness Act*; Technical Report; Illinois Department of Natural Resources: Springfield, IL, USA, 2015.

5.   Illinois Department of Natural Resources Office of Water Resources. *Model Stormwater Management Ordinance*; Illinois Department of Natural Resources: Springfield, IL, USA, 2015.

6.   Federal Emergency Management Agency (FEMA). *Hazus: FEMA's Methodology for Estimating Potential Losses from Disasters*; Federal Emergency Management Agency: Washington, DC, USA, 2018.

7.   Trenberth, K.E.; Dai, A.; Rasmussen, R.M.; Parsons, D.B. The Changing Character of Precipitation. *Bull. Am. Meteorol. Soc.* **2003**, *84*, 1205–1218. [CrossRef]

8.   Field, C.B.; Barros, V.; Stocker, T.F.; Qin, D.; Dokken, D.J.; Ebi, K.L.; Mastrandrea, M.D.; Mach, K.J.; Plattner, G.K.; Allen, S.K.; et al. *Managing the Risks of Extreme Events and Disasters to Advance Climate Change Adaptation: Special Report of the Intergovernmental Panel on Climate Change*; Cambridge University Press: Cambridge, UK, 2012.

9.   Milly, P.C.D.; Betancourt, J.; Falkenmark, M.; Hirsch, R.M.; Kundzewicz, Z.W.; Lettenmaier, D.P.; Stouffer, R.J. Stationarity Is Dead: Whither Water Management? *Science* **2008**, *319*, 573–574. [CrossRef]

10.   United Nations Framework Convention on Climate Change (UNFCCC). *Views and Information on the Effectiveness of the Nairobi Work Programme on Impacts, Vulnerability and Adaptation to Climate Change in Fulfilling Its Objective, Expected Outcome, Scope of Work and Modalities*; United Nations Framework Convention on Climate Change: Conn, Germany, 2010.

11.   Sussams, L.; Sheate, W.; Eales, R. Green infrastructure as a climate change adaptation policy intervention: Muddying the waters or clearing a path to a more secure future? *J. Environ. Manag.* **2015**, *147*, 184–193. [CrossRef]

12.   Mishra, V.; Lettenmaier, D.P. Climatic trends in major U.S. urban areas, 1950–2009. *Geophys. Res. Lett.* **2011**, *38*. [CrossRef]

13.   Allen, M.R.; Ingram, W.J. Constraints on future changes in climate and the hydrologic cycle. *Nature* **2002**, *419*, 224–232. [CrossRef]

14.   Karl, T.R.; Meehl, G.A.; Miller, C.D.; Hassol, S.J.; Waple, A.M.; Murray, W.L. *Weather and Climate Extremes in a Changing Climate*; Technical Report; U.S. Climate Change Science Program: Washington, DC, USA, 2008.

15.   Lenderink, G.; Van Meijgaard, E. Increase in hourly precipitation extremes beyond expectations from temperature changes. *Nat. Geosci.* **2008**, *1*, 511–514. [CrossRef]

16.   Mishra, V.; Dominguez, F.; Lettenmaier, D.P. Urban precipitation extremes: How reliable are regional climate models? *Geophys. Res. Lett.* **2012**, *39*. [CrossRef]

17.   Kunkel, K.E. North American Trends in Extreme Precipitation. *Nat. Hazards* **2003**, *29*, 291–305. [CrossRef]

18.   IPCC. *Climate Change 2013: The Physical Science Basis. Contribution of Working Group I to the Fifth Assessment Report of the Intergovernmental Panel on Climate Change*; Cambridge University Press: Cambridge, UK; New York, NY, USA, 2013; p. 1535. [CrossRef]

19.   Cantone, J.; Seo, Y.; Zimmer, A.; Schmidt, A.R.; Garcia, M.H. *Hydrologic Modeling of the Calumet TARP System*; Technical Report; Ven Te Chow Hydrosystems Laboratory, Department of Civil and Environmental Engineering, University of Illinois at Urbana-Champaign: Champaign, IL, USA, 2009.

20.   Cantone, J.; Schmidt, A.R. Improved understanding and prediction of the hydrologic response of highly urbanized catchments through development of the Illinois Urban Hydrologic Model. *Water Resour. Res.* **2011**, *47*. [CrossRef]

21.   Blöschl, G.; Sivapalan, M. Scale issues in hydrological modelling: A review. *Hydrol. Process.* **1995**, *9*, 251–290. [CrossRef]

22.   Seo, Y.; Schmidt, A.R. Network configuration and hydrograph sensitivity to storm kinematics. *Water Resour. Res.* **2013**, *49*, 1812–1827. [CrossRef]

23.   Hohenegger, C.; Brockhaus, P.; Schäer, C. Towards climate simulations at cloud-resolving scales. *Meteorol. Z.* **2008**, *17*, 383–394. [CrossRef]

24. Wakazuki, Y.; Nakamura, M.; Kanada, S.; Muroi, C. Climatological Reproducibility Evaluation and Future Climate Projection of Extreme Precipitation Events in the Baiu Season Using a High-Resolution Non-Hydrostatic RCM in Comparison with an AGCM. *J. Meteorol. Soc. Jpn. Ser. II* **2008**, *86*, 951–967. [CrossRef]

25. Gutowski, W.J.; Decker, S.G.; Donavon, R.A.; Pan, Z.; Arritt, R.W.; Takle, E.S. Temporal-Spatial Scales of Observed and Simulated Precipitation in Central U.S. Climate. *J. Clim.* **2003**, *16*, 3841–3847. [CrossRef]

26. Roberts, N.M.; Lean, H.W. Scale-Selective Verification of Rainfall Accumulations from High-Resolution Forecasts of Convective Events. *Mon. Weather Rev.* **2008**, *136*, 78–97. [CrossRef]

27. Maraun, D.; Wetterhall, F.; Ireson, A.M.; Chandler, R.E.; Kendon, E.J.; Widmann, M.; Brienen, S.; Rust, H.W.; Sauter, T.; Themeßl, M.; et al. Precipitation downscaling under climate change: Recent developments to bridge the gap between dynamical models and the end user. *Rev. Geophys.* **2010**, *48*. [CrossRef]

28. Markus, M.; Wuebbles, D.J.; Liang, X.Z.; Hayhoe, K.; Kristovich, D.A. Diagnostic analysis of future climate scenarios applied to urban flooding in the Chicago metropolitan area. *Clim. Chang.* **2012**, *111*, 879–902. [CrossRef]

29. Mearns, L.; McGinnis, S.; Arritt, R.; Biner, S.; Duffy, P.; Gutowski, W.; Held, I.; Jones, R.; Leung, R.; Nunes, A.; et al. The North American Regional Climate Change Assessment Program Dataset, 2007, updated 2014. Data downloaded 2016-12-04. Available online: https://www.narccap.ucar.edu/doc/pubs/bams-narccap-overview.pdf (accessed on 4 December 2016).

30. Taye, M.T.; Willems, P.; Block, P. Implications of climate change on hydrological extremes in the Blue Nile basin: A review. *J. Hydrol. Reg. Stud.* **2015**, *4*, 280–293. [CrossRef]

31. Wootten, A.; Terando, A.; Reich, B.J.; Boyles, R.P.; Semazzi, F. Characterizing Sources of Uncertainty from Global Climate Models and Downscaling Techniques. *J. Appl. Meteorol. Climatol.* **2017**, *56*, 3245–3262. [CrossRef]

32. Kendon, E.J.; Jones, R.G.; Kjellström, E.; Murphy, J.M. Using and Designing GCM-RCM Ensemble Regional Climate Projections. *J. Clim.* **2010**, *23*, 6485–6503. [CrossRef]

33. Kendon, E.J.; Rowell, D.P.; Jones, R.G. Mechanisms and reliability of future projected changes in daily precipitation. *Clim. Dyn.* **2010**, *35*, 489–509. [CrossRef]

34. Frei, C.; Schöll, R.; Fukutome, S.; Schmidli, J.; Vidale, P.L. Future change of precipitation extremes in Europe: Intercomparison of scenarios from regional climate models. *J. Geophys. Res. Atmos.* **2006**, *111*. [CrossRef]

35. Gutowski, W.J., Jr.; Giorgi, F.; Timbal, B.; Frigon, A.; Jacob, D.; Kang, H.S.; Raghavan, K.; Lee, B.; Lennard, C.; et al. WCRP COordinated Regional Downscaling EXperiment (CORDEX): A diagnostic MIP for CMIP6. *Geosci. Model Dev.* **2016**, *9*, 4087–4095. [CrossRef]

36. Kling, G.W.; Hayhoe, K.; Johnson, L.B.; Magnuson, J.J.; Polasky, S.; Robinson, S.K.; Shuter, B.J.; Wander, M.M.; Wuebbles, D.J.; Zak, D.R.; et al. *Confronting Climate Change in the Great Lakes Region: Impacts on Our Communities and Ecosystems*; Technical Report, Union of Concerned Scientists; Ecological Society of America: Washington, DC, USA, 2003.

37. Grenier, P.; Parent, A.C.; Huard, D.; Anctil, F.; Chaumont, D. An Assessment of Six Dissimilarity Metrics for Climate Analogs. *J. Appl. Meteorol. Climatol.* **2013**, *52*, 733–752. [CrossRef]

38. Retchless, D.P. Communicating climate change: Spatial analog versus color-banded isoline maps with and without accompanying text. *Cartogr. Geogr. Inf. Sci.* **2014**, *41*, 55–74. [CrossRef]

39. Kalkstein, L.S.; Greene, J.S. An evaluation of climate/mortality relationships in large US cities and the possible impacts of a climate change. *Environ. Health Perspect.* **1997**, *105*, 84–93. [CrossRef]

40. Hallegatte, S.; Hourcade, J.C.; Ambrosi, P. Using climate analogues for assessing climate change economic impacts in urban areas. *Clim. Chang.* **2007**, *82*, 47–60. [CrossRef]

41. Kopf, S.; Ha-Duong, M.; Hallegatte, S. Using maps of city analogues to display and interpret climate change scenarios and their uncertainty. *Nat. Hazards Earth Syst. Sci.* **2008**, *8*, 905–918. [CrossRef]

42. Kellett, J.; Hamilton, C.; Ness, D.; Pullen, S. Testing the limits of regional climate analogue studies: An Australian example. *Land Use Policy* **2015**, *44*, 54–61. [CrossRef]

43. Horváth, L.; Solymosi, N.; Gaál, M. Use of the spatial analogy to understand the effects of climate change. In *Environmental, Health and Humanity Issues in the Down Danubian Region*; World Scientific: Singapore, 2009; pp. 215–222. [CrossRef]

44. Ramírez-Villegas, J.; Lau, C.; Köhler, A.; Signer, J.; Jarvis, A.; Arnell, N.; Osborne, T.M.; Hooker, J. *Climate Analogues: Finding Tomorrow's Agriculture Today*; Technical Report, Working Paper 12; CCAFS: Copenhagen, Denmark, 2011.

45. Hirschboeck, K.K. Climate and floods. *US Geol. Surv. Water-Supply Pap.* **1991**, *2375*, 67–88.

46. Berne, A.; Delrieu, G.; Creutin, J.D.; Obled, C. Temporal and spatial resolution of rainfall measurements required for urban hydrology. *J. Hydrol.* **2004**, *299*, 166–179. [CrossRef]

47. Cristiano, E.; ten Veldhuis, M.C.; van de Giesen, N. Spatial and temporal variability of rainfall and their effects on hydrological response in urban areas—A review. *Hydrol. Earth Syst. Sci.* **2017**, *21*, 3859–3878. [CrossRef]

48. Hayhoe, K.; VanDorn, J.; Croley, T., II; Schlegal, N.; Wuebbles, D. Regional climate change projections for Chicago and the US Great Lakes. *J. Great Lakes Res.* **2010**, *36*, 7–21. [CrossRef]

49. Wuebbles, D.J.; Hayhoe, K.; Parzen, J. Introduction: Assessing the effects of climate change on Chicago and the Great Lakes. *J. Great Lakes Res.* **2010**, *36*, 1–6. [CrossRef]

50. Nakicenovic, N.; Alcamo, J.; Grubler, A.; Riahi, K.; Roehrl, R.; Rogner, H.H.; Victor, N. *Special Report on Emissions Scenarios (SRES), A Special Report of Working Group III of the Intergovernmental Panel on Climate Change*; Cambridge University Press: Cambridge, UK, 2000.

51. Markus, M.; Angel, J.; Byard, G.; McConkey, S.; Zhang, C.; Cai, X.; Notaro, M.; Ashfaq, M. Communicating the Impacts of Projected Climate Change on Heavy Rainfall Using a Weighted Ensemble Approach. *J. Hydrol. Eng.* **2018**, *23*, 04018004. [CrossRef]

52. Notaro, M.; Lorenz, D.; Hoving, C.; Schummer, M. Twenty-First-Century Projections of Snowfall and Winter Severity across Central-Eastern North America. *J. Clim.* **2014**, *27*, 6526–6550. [CrossRef]

53. Taylor, K.E.; Stouffer, R.J.; Meehl, G.A. An Overview of CMIP5 and the Experiment Design. *Bull. Am. Meteorol. Soc.* **2012**. [CrossRef]

54. Mearns, L.; McGinnis, S.; Korytina, D.; Arritt, R.; Biner, S.; Bukovsky, M.; Chang, H.I.; Christensen, O.; Herzmann, D.; Jiao, Y.; et al. *The NA-CORDEX Dataset*, version 1.0; 2017. Available online: https://doi.org/10.5065/D6SJ1JCH (accessed on 18 March 2019).

55. Wuebbles, D.J.; Meehl, G.; Hayhoe, K.; Karl, T.R.; Kunkel, K.; Santer, B.; Wehner, M.; Colle, B.; Fischer, E.M.; Fu, R.; et al. CMIP5 Climate Model Analyses: Climate Extremes in the United States. *Bull. Am. Meteorol. Soc.* **2014**. [CrossRef]

56. Meehl, G.A.; Covey, C.; Delworth, T.; Latif, M.; McAvaney, B.; Mitchell, J.F.B.; Stouffer, R.J.; Taylor, K.E. THE WCRP CMIP3 Multimodel Dataset: A New Era in Climate Change Research. *Bull. Am. Meteorol. Soc.* **2007**, *88*, 1383–1394. [CrossRef]

57. Metropolitan Water Reclamation District of Greater Chicago (MWRDGC). *Tunnel and Reservoir Plan (TARP)*; MWRDGC: Chicago, IL, USA, 1999.

58. Tang, Y.; Schmidt, A. Probabilistic Hydrologic Model to Simulate Response of Urban Drainage System to Implementation of Low Impact Development Stormwater Practices. In *World Environmental and Water Resources Congress 2013: Showcasing the Future*; American Society of Civil Engineers (ASCE): Reston, VA, USA, 2013. [CrossRef]

59. Wang, A.K.; Park, S.Y.; Huang, S.; Schmidt, A.R. Hydrologic Response of Sustainable Urban Drainage to Different Climate Scenarios. In *World Environmental and Water Resources Congress 2015: Floods, Droughts, and Ecosystems*; ASCE: Reston, VA, USA, 2015. [CrossRef]

60. Mearns, L.; Gutowski, W.; Jones, R.; Leung, R.; McGinnis, S.; Nunes, A.; Qian, Y. A Regional Climate Change Assessment Program for North America. *Eos Trans. Am. Geophys. Union* **2009**, *90*, 311. [CrossRef]

61. Bonnin, G.M.; Todd, D.; Lin, B.; Parzybok, T.; Yekta, M.; Riley, D. *Precipitation-Frequency Atlas of the United States, NOAA Atlas 14, Volume 2, Version 3.0*; Technical Report, NOAA, National Weather Service: Silver Spring, MD, USA, 2005.

62. Räisänen, J.; Ylhäisi, J.S. How Much Should Climate Model Output Be Smoothed in Space? *J. Clim.* **2011**, *24*, 867–880. [CrossRef]

63. Kopparla, P.; Fischer, E.M.; Hannay, C.; Knutti, R. Improved simulation of extreme precipitation in a high-resolution atmosphere model. *Geophys. Res. Lett.* **2013**, *40*, 5803–5808. [CrossRef]

64. CDO. *Climate Data Operators*; 2018. Available online: https://code.mpimet.mpg.de/projects/cdo (accessed on 16 May 2019).

65. Hosking, J.R.M.; Wallis, J.R. *Regional Frequency Analysis: An Approach Based on L-Moments*; Cambridge University Press: Cambridge, UK, 1997. [CrossRef]

66. Greenwood, J.A.; Landwehr, J.M.; Matalas, N.C.; Wallis, J.R. Probability weighted moments: Definition and relation to parameters of several distributions expressable in inverse form. *Water Resour. Res.* **1979**, *15*, 1049–1054. [CrossRef]

67. Hosking, J.R.M. *Package 'lmom'*; R Package, Version 2.6; 2017. Available online: https://cran.r-project.org/package=lmom (access on 18 March 2019).

68. Papalexiou, S.M.; Koutsoyiannis, D. Battle of extreme value distributions: A global survey on extreme daily rainfall. *Water Resour. Res.* **2013**, *49*, 187–201. [CrossRef]

69. Serinaldi, F.; Kilsby, C.G. Rainfall extremes: Toward reconciliation after the battle of distributions. *Water Resour. Res.* **2014**, *50*, 336–352. [CrossRef] [PubMed]

70. Zalina, M.; Desa, M.; Nguyen, V.T.A.; Kassim, A. Selecting a probability distribution for extreme rainfall series in Malaysia. *Water Sci. Technol.* **2002**, *45*, 63–68. [CrossRef]

71. Anderson, B.; Siriwardena, L.; Western, A.; Chiew, F.; Seed, A.; Bloschl, G. Which theoretical distribution function best fits measured within day rainfall distributions across Australia? In *30th Hydrology & Water Resources Symposium: Past, Present & Future. Conference Design*; Institution of Engineers, Australia: Launceston, Australia, 2006; pp. 498–503.

72. Alam, M.A.; Emura, K.; Farnham, C.; Yuan, J. Best-Fit Probability Distributions and Return Periods for Maximum Monthly Rainfall in Bangladesh. *Climate* **2018**, *6*, 9. [CrossRef]

73. Chow, V.T.; Maidment, D.R.; Mays, L.W. *Applied Hydrology*; McGraw-Hill International Editions: New York, NY, USA, 1988.

74. Whetton, P.; Macadam, I.; Bathols, J.; O'Grady, J. Assessment of the use of current climate patterns to evaluate regional enhanced greenhouse response patterns of climate models. *Geophys. Res. Lett.* **2007**, *34*. [CrossRef]

75. Reifen, C.; Toumi, R. Climate projections: Past performance no guarantee of future skill? *Geophys. Res. Lett.* **2009**, *36*. [CrossRef]

76. Hayhoe, K.; Edmonds, J.; Kopp, R.; LeGrande, A.; Sanderson, B.; Wehner, M.; Wuebbles, D. Climate models, scenarios, and projections. In *Climate Science Special Report: Fourth National Climate Assessment, Volume I*; Wuebbles, D., Fahey, D., Hibbard, K., Dokken, D., Stewart, B., Maycock, T., Eds.; U.S. Global Change Research Program: Washington, DC, USA, 2017; Chapter 4, pp. 133–160. [CrossRef]

77. Anandhi, A.; Frei, A.; Pierson, D.C.; Schneiderman, E.M.; Zion, M.S.; Lounsbury, D.; Matonse, A.H. Examination of change factor methodologies for climate change impact assessment. *Water Resour. Res.* **2011**, *47*. [CrossRef]

78. Lafon, T.; Dadson, S.; Buys, G.; Prudhomme, C. Bias correction of daily precipitation simulated by a regional climate model: A comparison of methods. *Int. J. Climatol.* **2013**, *33*, 1367–1381. [CrossRef]

79. Monette, A.; Sushama, L.; Khaliq, M.N.; Laprise, R.; Roy, R. Projected changes to precipitation extremes for northeast Canadian watersheds using a multi-RCM ensemble. *J. Geophys. Res. Atmos.* **2012**, *117*. [CrossRef]

80. Khaliq, M.; Sushama, L.; Monette, A.; Wheater, H. Seasonal and extreme precipitation characteristics for the watersheds of the Canadian Prairie Provinces as simulated by the NARCCAP multi-RCM ensemble. *Clim. Dyn.* **2015**, *44*, 255–277. [CrossRef]

81. Kunkel, K.E.; Easterling, D.R.; Kristovich, D.A.R.; Gleason, B.; Stoecker, L.; Smith, R. Meteorological Causes of the Secular Variations in Observed Extreme Precipitation Events for the Conterminous United States. *J. Hydrometeorol.* **2012**, *13*, 1131–1141. [CrossRef]

82. Feng, Z.; Leung, L.R.; Hagos, S.; Houze, R.A.; Burleyson, C.D.; Balaguru, K. More frequent intense and long-lived storms dominate the springtime trend in central US rainfall. *Nat. Commun.* **2016**, *7*, 13429. [CrossRef]

water MDPI

Article

A Study on Climate-Driven Flash Flood Risks in the Boise River Watershed, Idaho

Jae Hyeon Ryu [1],* and Jungjin Kim [2]

[1] Department of Soil and Water Systems, University of Idaho, 322E. Front ST, Boise, ID 83702, USA
[2] Texas A&M AgriLife Research (Texas A&M University System), P.O. Box 1658, Vernon, TX 76384, USA;
     jungjin.kim@ag.tamu.edu
*   Correspondence: jryu@uidaho.edu

Received: 15 March 2019; Accepted: 14 May 2019; Published: 18 May 2019

Abstract: We conducted a study on climate-driven flash flood risk in the Boise River Watershed using flood frequency analysis and climate-driven hydrological simulations over the next few decades. Three different distribution families, including the Gumbel Extreme Value Type I (GEV), the 3-parameter log-normal (LN3) and log-Pearson type III (LP3) are used to explore the likelihood of potential flash flood based on the 3-day running total streamflow sequences (3D flows). Climate-driven ensemble streamflows are also generated to evaluate how future climate variability affects local hydrology associated with potential flash flood risks. The result indicates that future climate change and variability may contribute to potential flash floods in the study area, but incorporating embedded-uncertainties inherited from climate models into water resource planning would be still challenging because grand investments are necessary to mitigate such risks within institutional and community consensus. Nonetheless, this study will provide useful insights for water managers to plan out sustainable water resources management under an uncertain and changing climate.

Keywords: flood frequency analysis; flash flood; climate change and variability; Boise River Watershed; HSPF

1. Introduction

Climate variability and change continues to increase the risk and frequency of floods for inland communities in the United States (US) [1–4]. Floods in 2017 alone claimed more than 3 billion dollars in property damages and crop losses [5]. As global warming shifts rainfall patters, more frequent heavy rain is likely contributing to flash floods at the urban-rural interface, such as the Boise River Watershed (BRW) [6]. In general, snowmelt-streamflow dominates high volume in many western watersheds during spring and summer [7,8]. Thus, heavy snowfall and accumulation in winter can elevate potential risks of flash flooding during snow-melting season. Over the last few years, this consequence of heavy snowfall often affects streamflow augmentation in the Boise River so that the second highest inflows to reservoirs upstream is recorded in water year 2017 (October 2016–September 2017) [9]. Such a high-volume water condition began increasing management concerns for reservoir operators and homeowners who live in the flood plain.

Recent studies show that the global climate cycle will create and intensify more severe frequent floods in many regions, resulting in threats to the reliability and resiliency of water resources infrastructure [10,11]. Many previous studies have investigated long-term hydrologic variability associated with climate change [12–15]. The general circulation models (GCMs) are commonly used to characterize local hydrologic conditions induced by climate variability and change over the next few decades. For instance, because of the timing change of snowfall and snowmelt in the western states, regional water resources management is increasingly facing additional challenges; thus, heavy snowfall increases potential risks of flash flood in the snow-dominated watershed. Floods may also intensify in many regions where total precipitation is even projected to decline due to climate uncertainties [14–16]. Based on the evidence of a larger proportion of snowmelt-driven streamflow volume during springtime leveraged by temperature increase, potential impacts of climate change on streamflow in the western states are likely increasing [12,17].

Many previous studies, however, focused on hydrologic consequence of climate change scenarios using statistical downscaling and bias correction processes [13,18,19]. Thus, given the dominantly linear response of the GCMs, future perturbations of hydrologic cycles induced by climate change were investigated to characterize climate-induced hydrological impacts at the regional scales. Relatively little study has been done to explore the risk of potential flash floods associated with climate variability using frequency analysis [20].

In this study, therefore, we investigate how future climate variability can characterize potential flash flood risks in the Boise River Watershed. Using both flood frequency analysis and future ensemble streamflow generations with climate inputs, potential flash flood events are analyzed. We anticipate that the result from this study will provide useful insights for local water managers to plan out future flood mitigation strategies in a changing global environment.

2. Study Area

The Boise River Watershed (BRW) is selected as the study area (Figure 1). As a tributary of the Snake River system, the BRW plays a key role of providing water to Boise metropolitan areas, including Boise, Nampa, Meridian, and Caldwell. The drainage area of the basin is about 10,619 km² with a mainstream length of 164 km stretch and flows into the Snake River near Parma. More than 40% of Idaho residents live in this basin and 60% of people of that are residing around the floodplain [21]. The main physical and geographic characteristic of the BRW is a greater proportion of precipitation falling at higher elevations. It becomes the cause of predictably high flows due to the snow melting process so that the localized flood event is often observed during late spring and early summer.

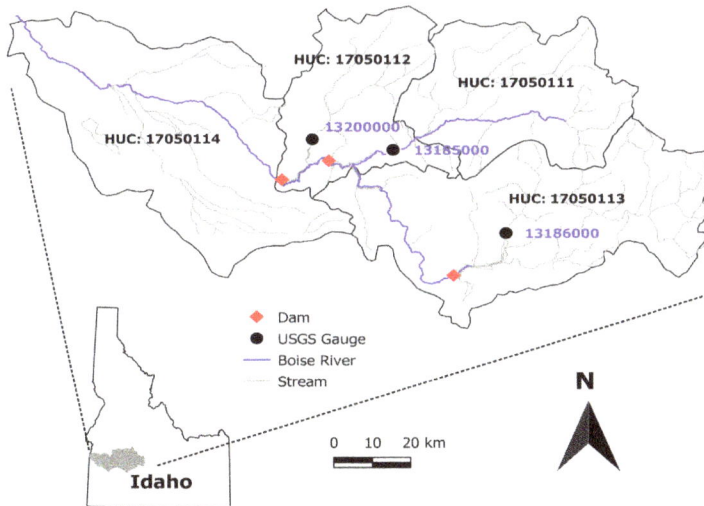

Figure 1. Map of the Boise River Watershed.

The recent flash flood induced by heavy snowfall 2017 further highlights a research proposal to increase water storage capacity of the Boise River system by raising small-portion elevation of the existing dams, including Lucky Peak, Arrowrock and Anderson Ranch. The Bureau of Reclamation is currently conducting the feasibility study of the dams under the December 2016 Federal Water Infrastructure Improvements for the Nation Act, which may also authorize funding for construction of projects by 1 January 2021 [22]. Additional water capacity in the BRW (if this project is complete) will provide more flexibility for water managers to mitigate impacts driven by climate-induced hydro extremes (flood and drought). Seasonal streamflow for three stations managed by United States Geological Survey (USGS), including USGS: 13200000 (OBS1), 13185000 (OBS2) and 13186000 (OBS3) are observed. As shown in Figure 2, the seasonal trends at these stations are distinct in the sense that snow-melting streamflows are dominant during summer, while rainfalls in later fall is also contributing to streamflow before major snowfall starts.

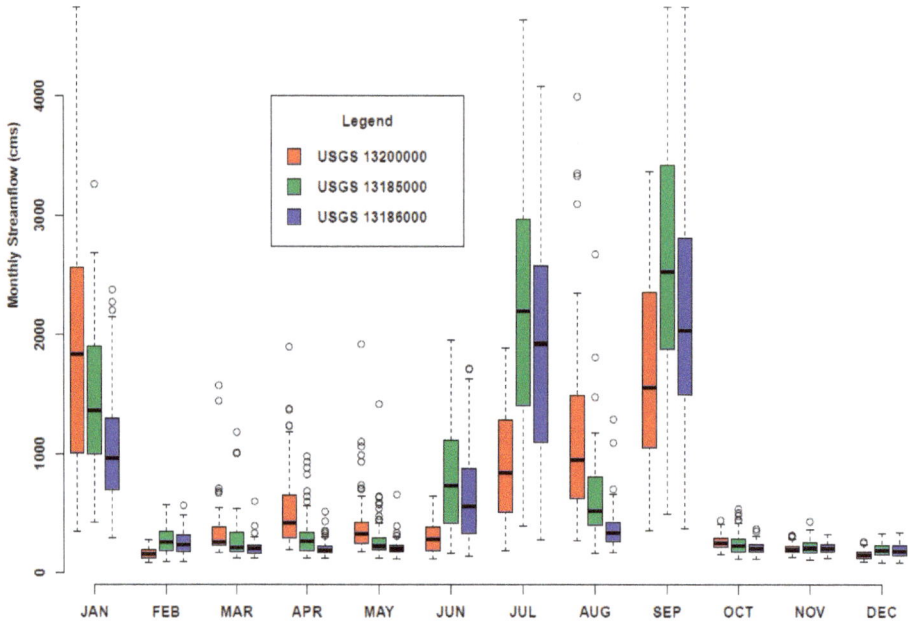

Figure 2. Box plots of the observed seasonal streamflow at the selected United States Geological Survey (USGS) stations (OBS1: USGS 1320000, OBS2: USGS 13185000, OBS3: USGS 13186000).

3. Methodology

3.1. Flash Flood Frequency

For flood frequency analysis, the magnitudes of a single hydro variable, such as annual maximum flood peak is widely used in hydro communities. For this study, 3-day running total streamflow sequences (3D flows) was utilized to better represent potential flash floods. Since a flash flood is caused by heavy rain and/or snowmelt streamflow in a short period of time, the maxim value of 3D flow at the given month was selected to consider independent and identically distributed variants (iid) for frequency analysis. For example, the flash flood in 2017 at OBS2 is recorded 876.69 cubic meter per second (cms), which is the second highest flow (7 May 2017) after 904.44 cms (27 April 2012) (see Table 1).

Table 1. The 3-day running total streamflows at the selected USGS stations (OBS1: USGS 13200000, OBS2: USGS 13185000, OBS3: USGS 13186000).

Index	OBS1		OBS2		OBS3	
	Date	Flow	Date	Flow	Date	Flow
1	7 April 1951	179.53	28 May 1951	551.33	28 May 1951	420.79
2	27 April 1952	282.88	27April 1952	686.97	4 May1952	430.42
3	28 April 1953	133.37	13 June 1953	626.08	13June 1953	352.26
4	18 April 1954	121.48	20 May 1954	625.80	20 May 1954	408.89
5	23 December 1955	292.23	23 December 1955	575.68	10 June 1955	306.67
6	16 April 1956	189.16	24 May 1956	857.43	24 May 1956	592.67
7	30 May 1970	149.23	5 June 1957	663.75	5 June 1957	473.74
8	18 April 1958	162.26	21 May 1958	818.07	22 May 1958	609.94
9	6 April 1959	90.61	14 June 1959	390.77	14 June 1959	235.31
10	7 April 1960	150.65	12 May 1960	458.73	12 May 1960	318.85
11	4 April 1961	53.43	26 May 1961	364.72	26 May 1961	216.34
12	19 April 1970	108.74	20 April 1962	393.60	12 June 1962	281.19
13	7 April 1963	73.14	24 May 1963	453.35	24 May 1963	326.21
14	24 December 1964	305.26	24 December 1964	777.30	21 May 1964	281.19
15	23 April 1965	325.36	11 June 1965	682.43	11 June 1965	550.76
16	1 April 1966	64.34	8 May 1966	321.11	9 May 1966	233.05
17	23 May 1967	72.69	23 May 1967	577.66	24 May 1967	466.09
18	23 February 1968	74.39	4 June 1968	295.63	4 June 1968	180.09
19	6 April 1969	205.30	14 May 1969	543.12	14 May 1969	480.54
20	24 May 1970	103.92	26 May 1970	569.45	8 June 1970	382.84
21	5 May 1971	185.76	14 May 1971	667.71	13 May 1971	518.48
22	19 March 1972	188.59	2 June 1972	784.94	9 June 1972	510.27
23	15 April 1973	54.45	19 May 1973	435.80	19 May 1973	257.40
24	31 March 1974	206.15	16 June 1974	805.33	16 June 1974	485.35
25	16 May 1975	225.68	16 May 1975	627.50	7 June 1975	467.51
26	10 April 1976	156.31	12 May 1976	527.26	15 May 1976	335.27
27	16 December 1977	76.88	16 December 1977	208.13	10 June 1977	60.37
28	31 March 1978	159.99	9 June 1978	496.11	9 June 1978	358.21
29	17 May 1979	48.85	25 May 1979	406.63	25 May 1979	266.18
30	24 April 1980	138.75	6 May 1980	491.01	6 May 1980	334.14
31	21 April 1981	65.69	9 June 1981	413.14	9 June 1981	240.41
32	14 April 1982	212.94	25 May 1982	633.45	18 June 1982	503.19
33	13 March 1983	257.97	29 May 1983	871.59	29 May 1983	643.36
34	18 April 1984	196.80	15 May 1984	711.60	15 May 1984	496.96
35	11 April 1985	120.91	4 May 1985	332.72	25 May 1985	244.09
36	24 February 1986	253.44	31 May 1986	768.23	31 May 1986	557.27
37	14 March 1987	47.91	30 April 1987	242.39	30 April 1987	146.40
38	5 April 1988	41.00	25 May 1988	260.23	25 May 1988	177.26
39	20 April 1989	155.74	10 May 1989	466.38	10 May 1989	342.35
40	29 April 1990	98.00	29 April 1990	280.90	31 May 1990	167.92
41	18 May 1991	34.26	4 June 1991	258.53	12 June 1991	179.53
42	22 February 1992	36.10	8 May 1992	193.12	8 May 1992	116.10
43	5 April 1993	167.64	15 May 1993	675.36	21 May 1993	390.49
44	22 April 1994	28.57	12 May 1994	235.03	13 May 1994	137.62
45	8 April 1995	150.36	4 June 1995	518.20	4 June 1995	425.32
46	31 December 1996	152.88	16 May 1996	790.89	17 May 1996	552.74
47	2 January 1997	301.29	16 May 1997	856.02	17 May 1997	656.10
48	28 May 1998	169.33	27 May 1998	468.64	10 May 1998	312.62
49	20 April 1999	158.29	26 May 1999	657.23	26 May 1999	438.06
50	14 April 2000	90.73	24 May 2000	387.37	24 May 2000	255.42
51	25 March 2001	34.15	16 May 2001	273.26	16 May 2001	140.45
52	15 April 2002	157.72	15 April 2002	479.12	1 June 2002	280.34
53	27 March 2003	67.42	30 May 2003	689.23	30 May 2003	467.51
54	7 April 2004	99.39	5 June 2004	284.58	6 May 2004	171.03

Table 1. *Cont.*

Index	OBS1		OBS2		OBS3	
	Date	Flow	Date	Flow	Date	Flow
55	20 May 2005	59.81	20 May 2005	477.14	20 May 2005	381.99
56	6 April 2006	293.93	20 May 2006	844.69	20 May 2006	651.85
57	14 March 2007	57.14	2 May 2007	291.38	13 May 2007	152.63
58	20 May 2008	105.62	20 May 2008	760.02	20 May 2008	412.86
59	22 April 2009	96.56	20 May 2009	508.85	1 June 2009	325.36
60	6 June 2010	103.07	6 June 2010	726.04	6 June 2010	413.99
61	18 April 2011	152.91	15 May 2011	792.30	15 May 2011	416.82
62	1 April 2012	173.87	27 April 2012	904.44	26 April 2012	538.02
63	7 April 2013	34.77	14 May 2013	365.29	14 May 2013	220.02
64	11 March 2014	75.69	27 May 2014	489.88	27 May 2014	274.39
65	10 February 2015	117.80	9 February 2015	303.56	26 May 2015	160.84
66	14 March 2016	99.68	13 April 2016	439.76	13 April 2016	298.46
67	21 March 2017	318.28	7 May 2017	876.69	7 May 2017	813.54

Figure 3 illustrates the number of occurrences of 3D flows each month starting from January 1951 to December 2017 at the three USGS stations (OBS1, OBS2, and OBS3). It appears that the likelihood of maximum 3D flows at the given month is noticeably observed in April and May at both OBS2 and OBS3, while such flow is also observed in March at OBS1.

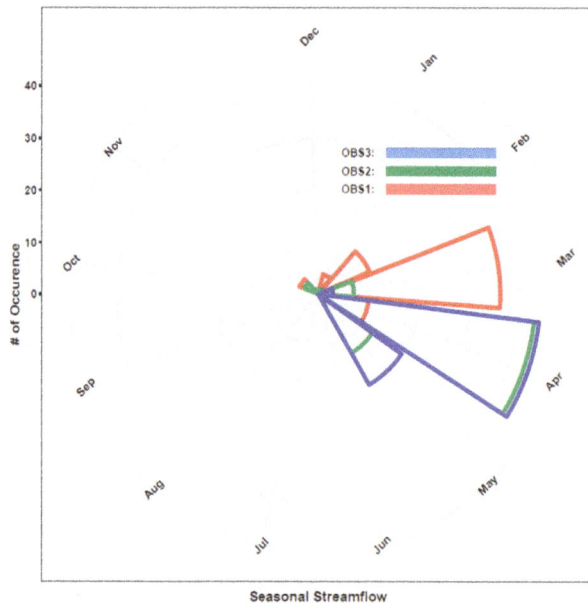

Figure 3. The number of occurrences of the maximum 3-day running total streamflow sequences (3D flows) at the given year for January 1 1951 to December 31 2017 at the selected USGS stations (OBS1: USGS 1320000, OBS2: USGS 13185000, OBS3: USGS 13186000).

Three distribution families, including the generalized extreme value type I (GEV), the 3-parameter lognormal (LN3) and Pearson distributions (LP3) [23–25] are commonly used for flood frequency analysis. The parameters of these distributions, however, should be estimated from several statistical methods, but the method of moment (MOM) was selected for the curve fitting based on the previous research [26]. For GEV, the reduced extreme value variate, X_i, can be defined as a function of the Weibull plotting position, q_i, which is the probability of the ith-largest event from the sample size, n.

Thus, the points when plotted would apart from sampling fluctuation, lie on a straight line through the original [27].

$$q_i = \frac{i}{n+1},\tag{1}$$

$$X_i = -\ln[-\ln(1-q_i)],\tag{2}$$

where, ln is the natural logarithm [28,29]. The specified position of a ith-flood, Y_i, can be defined as [30]:

$$Y_i = \overline{Y} + K_i \sigma_Y,\tag{3}$$

where, $\overline{Y}$ is the mean of the flood series, σ_y is the standard deviation of the series, and K_i is a frequency factor defined by a specific distribution, which is GEV I (GEV) in this case [27,31,32].

$$K_i = (0.7797X_i - 0.45).\tag{4}$$

In order to plot the fitted values from three-parameter lognormal distribution, mean, standard deviation, and location parameter should be estimated [33]. The parameter estimation for the location parameter, in particular, is more difficult in the sense that an iterative solution of a nonlinear equation should be achieved to retain their desirable asymptotic properties. [34]. The method of quantiles would be a feasible solution to estimate the location parameter, τ.

$$\tau = \frac{x_q x_{1-q} - (x_{0.5})^2}{x_q + x_{1-q} - 2x_{0.5}},\tag{5}$$

where, x_q, x_{1-q}, and $x_{0.5}$ are the largest, smallest, and median of the observations. This choice of the values ensures that the estimated lower bound is smaller than the smallest observation so that the fitted lower bound is reasonable [34]. For the three-parameter log normal distribution, Y_i may be written:

$$Y_i = \tau + \exp(a + bq_i),\tag{6}$$

$$a = \frac{1}{n}\sum_{i=1}^{n}\ln(x_i - \tau),\tag{7}$$

$$b = \sqrt{\frac{\sum_{i=1}^{N}(x_i - \tau)^2}{N-1}}.\tag{8}$$

Researchers [35] demonstrate parameter estimation to generate a sample from a log Pearson type 3 distribution (LP3). The probability density function of LP3 can be represented as:

$$f(x) = \frac{\lambda^\beta (x-\zeta)^{\beta-1}\exp(-\lambda(x-\zeta))}{\Gamma(\beta)},\tag{9}$$

where, λ, β and ζ are parameters for LP3 and the method of moment is applied for parameter estimation [28].

3.2. Hydrological Model Used

Hydrological Simulation Program FORTRAN (HSPF) was used as a hydrological model to simulate the past and future hydrological consequences associated with climate variability [36–38]. HSPF is a process-based, river basin-scale, and semi-distributed model that simulates hydrological conditions through Hydrological Response Units (HRUs) within the watershed. Built upon Sandford Watershed Model IV [39,40], HSPF is widely used for water quantity and quality simulations for many national and international watersheds [41–45]. For hydrological simulation, a series of datasets was used, including the Digital Elevation Model (DEM) in 30-meter resolution and the National

Hydrography Dataset (NHD). As environmental background data, the 2011 Land Use Land Cover (LULC) datasets provided by National Land Cover Database (NLCD) were used to perform a more detailed assessment of current LULC conditions in three watersheds. For climate forcing data, phase 2 of the North American Land Data Assimilation System (NLDAS-2) data, including precipitation, temperature, and potential evapotranspiration (PET) at an hourly time step were used [46]. NLDAS-2 is in 1/8th-degree grid spacing (about 12 × 12 km) and the simulation period is set for 1 January 1979 through 31 December 2015 at an hourly time step.

For HSPF calibration and validation, we utilized observed daily streamflow for calibration (1979–2005) and validation (2006–2015). Initial 2-year simulations (1979–1980) were used as the warm up period. A total of three observed streamflow stations located in above reservoirs were selected for calibration target points because these stations are less influenced by anthropogenic water activities (e.g., diversion, irrigation, and dam operations) (see Figure 1). A model-independent parameter estimation package (PEST) was used as an automatic calibration tool in BEOPEST environment, which is a special version of PEST in parallel computing to save calibration time and to improve model performance. Model performance was measured based on criteria, including correlation coefficient (R), the Nash–Sutcliffe efficiency (NSE), observation standard deviation ratio (RSR), and percentage of bias (PBIAS), which are typically used as described in the Appendix A. The more detailed HSPF modeling and calibration efforts can be found in the literature [13].

3.3. Future Climate Scenarios Implemented

A total of 13 Global Circulation Models (GCMs) under representative concentration pathways (RCPs), including mid-range mitigation emission scenarios (RCP4.5) and high emission scenarios (RCP8.5) were used to generate climate-driven streamflows over the next few decades until 31 December 2099. Using Multivariate Adaptive Constructed Analogs (MACA)-based Coupled Model Inter-Comparison Project (CMIP5) statistically downscaled data for conterminous USA [47], the extended future streamflows were generated at the selected USGS stations (OBS1, OBS2 and OBS3). There were a total of 13 MACA. More detailed information about the GCMs are listed in Table 2.

Table 2. List of the Coupled Model Inter-Comparison Project (CMIP5) models used in this study.

Model	Modeling Group	Note
BCC-CSM1-1	Beijing Climate Center, China Meteorological Administration, China	
BCC-CSM1-1m		
BNU-ESM	College of Global Change and Earth System Science, Beijing Normal University, China	
CANESM2	Canadian Centre for Climate Modelling and Analysis, Canada	
CCSM4	National Center for Atmospheric Research, USA	
CNRM-CM5	Centre National de Recherches Meteorologiques, Meteo-France, France	1. 4 km spatial resolution
CSIRO-MK3	Commonwealth Scientific and Industrial Research Organisation in collaboration with the Queensland Climate Change Centre of Excellence, Australia	2. Scenario: RCP4.5, RCP8.5
GFDL-ESM2G	NOAA Geophysical Fluid Dynamics Laboratory (GFDL), USA	
IPSL-CM5A-LR		
IPSL-CM5A-MR	Institute Pierre-Simon Laplace, France	
IPSL-CM5B-LR		
MIROC5	Atmosphere and Ocean Research Institute, Japan	
MIROC-ESM	Japan Agency for Marine-Earth Science and Technology, Japan	
MIROC-ESM-CHEM		

Basically, RCPs indicate the estimation of the radiative forcing associated with future climate variability and change. For example, RCP8.5 represents the increase of the radiative forcing throughout the 21st century before it reaches a level to 8.5 W/m^2 at the end of the century. All datasets covering the period 1979–2099 were obtained from [47]. Although future GCM data would be useful, additional efforts are needed to incorporate such data into HSPF modeling framework. Thus, bias correction was applied using a quantile-based mapping technique associated with the synthetic gamma distribution function to cross-validate GCMs and NLDAS-2 dataset. The bias correction assumes the biases represents the same pattern in both present and future climate conditions. It was based on the comparison between Cumulative Distribution Function (CDF) for NLDAS-2 and GCM data within the same time window. Thus, the bias between the GCM and NLDAS-2 during the reference period (1979–2005) was also considered to adjust future climate conditions prior to HSPF simulations as forcing inputs. The CDF was first calculated based on the month-specific probability distribution for monthly GCM and NLDAS-2 data, including precipitation, temperature and potential evapotranspiration (PET). The inverse CDF of the gamma function was then used to apply bias correction for GCMs from NLDAS-2. The more detailed process can be found at [13].

4. Results

Figures 4–6 illustrate a comparison of the 3D flows against the Gumbel reduced variable for the selected USGS OBS1, OB2, and OBS3, respectively. Simple correlation coefficients and Kolmogorov–Smirnov statistic were computed for goodness-of-fit and it is concluded that all three methods are acceptable because the correlation coefficient is high enough (>0.98) and the Kolmogorov–Smimov empirical statistic [48], Dn (Dn = 0.16) is smaller with 95% confidence level. The interested reader may also apply another goodness of fit, such as chi square test [49] for cross validation, when necessary. Confidence limits suggested by [50] were also applied to provide useful insights for water managers, who may utilize this information to mitigate impacts driven by flash floods. Note that the upper and lower bound lines are plotted based on GEV and those lines indicate a wide range of uncertainty for GEV Type I distribution at the 95% confidence level.

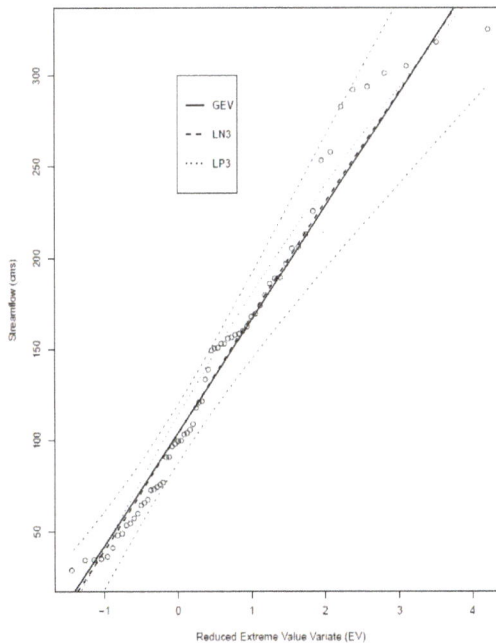

Figure 4. Comparison of three theoretical distributions (Gumbel Extreme Value Type I (GEV), 3-parameter log-normal (LN3), log-Pearson type III (LP3)) for annual 3D flow frequency at the 95% confidence level at OBS1 (USGS 13200000).

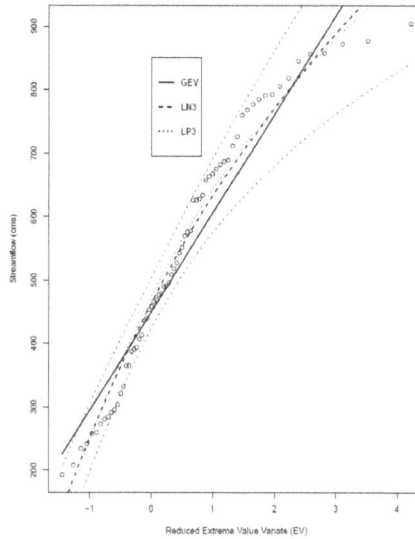

Figure 5. Comparison of three theoretical distributions (GEV, LN3, LP3) for annual 3D flow frequency at the 95% confidence level at OBS2 (USGS 13185000).

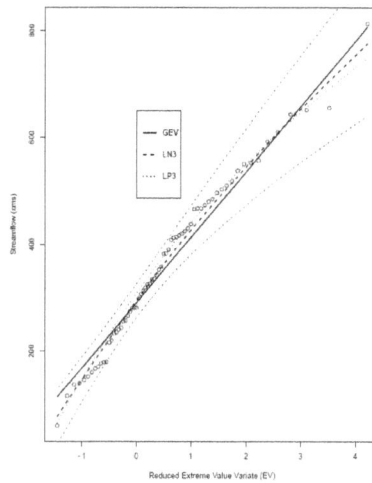

Figure 6. Comparison of three theoretical distributions (GEV, LN3, LP3) for annual 3D flow frequency at the 95% confidence level at OBS3 (USGS 13186000).

The Monte Carlo simulation was also conducted to understand the impact of risk and uncertainty in flash flood events. A total of 1000 streamflow sequences were generated and distinct 30, 60 and 90 samples were selected to observe a 95% confidence level. Table 3 shows the 3D peak flow from Monte Carlo simulation associated with different return periods (25, 50, 100, 150 and 200 years) based on Gumbel Extreme Value Type I (GEV). Note that the return period of 200 years can be interpreted as the total span of streamflow data in BRW has 200-year records from 1951 to 2150 (200 years), which is beyond of the climate model projection until 2099.

Table 3. The 3D peak flows from Monte Carlo simulation from 1000 streamflow sequences with different sample sizes (30, 60, 90) and return periods (25, 50, 100, and 200 years) based on Gumbel Extreme Value Type I (GEV).

OBS1		25	50	100	150	200
N = 30	Upper	348	410	458	486	514
	Lower	255	285	319	337	353
N = 60	Upper	336	390	437	464	491
	Lower	268	302	341	358	380
N = 90	Upper	333	381	426	458	480
	Lower	274	313	352	374	386
OBS2		25	50	100	150	200
N = 30	Upper	1075	1206	1337	1425	1471
	Lower	822	918	983	1039	1067
N = 60	Upper	1033	1166	1278	1371	1422
	Lower	858	952	1044	1093	1123
N = 90	Upper	1025	1143	1268	1337	1402
	Lower	876	972	1062	1118	1164
OBS3		25	50	100	150	200
N = 30	Upper	780	884	985	1076	1109
	Lower	587	654	714	761	780
N = 60	Upper	753	854	950	1013	1053
	Lower	612	686	759	805	829
N = 90	Upper	739	842	929	994	1029
	Lower	628	705	781	822	844

The streamflow calibration and validation were also performed to generate climate-induced future streamflows at BRW. The calibration and validation periods of streamflow are 1979–2005 (27 years) and 2006–2015 (10 years), respectively, but the first two years (1979–1980) were used as a warm up period. Table 4 shows the calibration and validation results for performance measures of streamflow at BRW using daily and monthly time steps. Based on criteria and recommended statistics (see Appendix A) for model performances [51,52], all three observed stations, OBS1, OBS2 and OBS3 show good model performance (e.g., R^2 = 0.87, NS = 0.86, and RSR = 0.37, and PBIAS = 11.10 at OBS1) during the calibration period. Overall, the calibrated HSPF performs very well to generate climate-driven future streamflows with GCMs inputs.

Table 5 lists the maximum of climate-driven ensemble streamflows (3D flows) from HSPF simulations with GCMs inputs. Both RCP 4.5 and RCP 8.5 scenarios are incorporated into HSPF to explore potential flood risks over the next few decades. It appears that RCP 4.5-induced streamflows might not have a great influence on the difference in the overall 3D flows at the selected stations. However, when the RCP 8.5 scenario was used, the significant increase was observed at OBS2 and OBS3. Based on the flood frequency analysis, the maximum of 3D flows at OBS2 and OBS3 are reported 1471 cms (N = 30) and 1109 cms (N = 3), respectively, which is much less than that from HSPF with GCMs inputs (see Table 5). This implies that uncertainties embedded in GCMs is quite large as opposed to the hydro stationarity—the idea that natural systems fluctuate within an unchanging envelop of historic flow variability [53–55]. Such an uncertainty, perhaps, can be reduced through more cohesive joint modeling efforts from the field of climatology and hydrology. Thus, the regional climate models are evolving with additional information and new approaches to better increase the predictability using any large-scale driving data, including aerosols and chemical species [56]. Additionally, the fast-moving technologies and applications, such as high-performance computing, computer parallelism in hydrological modeling [57], and unmanned aerial system (UAS) for flood mapping would be another avenue to improve predictability by mitigating uncertainty and risks associated with other foreseen factors [13] (e.g., population growth, urbanization, and economic development).

Table 4. Performance statistics for the calibrated (1979–2005) and validated (2006–2015) monthly streamflow at the Boise River Watershed using daily and monthly time steps.

Variable		OBS1		OBS2		OBS3	
		Cal	Val	Cal	Val	Cal	Val
R^2	Daily	0.82	0.72	0.78	0.74	0.81	0.87
	Monthly	0.87	0.81	0.85	0.80	0.85	0.92
NS	Daily	0.81	0.70	0.77	0.73	0.79	0.86
	Monthly	0.86	0.87	0.85	0.89	0.84	0.95
RSR	Daily	0.43	0.54	0.48	0.52	0.46	0.37
	Monthly	0.37	0.36	0.39	0.34	0.40	0.22
PBIAS (%)	Daily	11.11	17.35	7.82	3.19	9.74	1.50
	Monthly	11.10	17.41	7.77	3.18	9.79	1.64

Table 5. The maximum of 3D flow from Hydrological Simulation Program FORTRAN (HSPF) simulations with Global Circulation Models (GCMs) inputs.

Climate Scenario	USGS Station	Streamflow	Date	Climate Model
RCP 4.5	OBS1	985.83	30 December 2011	Ipsl.cm5a
	OBS2	2469.16	30 December 2011	Ipsl.cm5a
	OBS3	1777.35	8 February 2015	Bcc.scm1
RCP 8.5	OBS1	776.65	16 March 1998	Ipsl.cm5b
	OBS2	1636.52	9 January 2089	Canesm2
	OBS3	2563.15	18 January 2089	Canesm2

For example, Figures 7–9 illustrate the time series of ensemble 3D flows at OBS1, OBS2 and OBS3 respectively from HSPF associated with each of the climate projections. Note that logarithm base 10 is applied to the flow to show general trends of the flow over the next few decades until 2099. One can see that the magnitude of the projected annual 3D peaks varies in different ways for every projection. These peaks would correspond to flash flood values with a return period greater than 140 years when compared to historic observation (1951–2017, 67 years). The linear regression model was then applied to draw a trend line with 95% confidence levels for visual inspection. Additionally, the upper and lower envelop lines indicating 85% and 25% of 3D flows are drawn to provide a general insight for the reader. Unlike 3D flows at OBS1 and OBS2, the climate-driven 3D flows at OBS3 shows an increasing trend with 95% confidence. However, overall climate-driven 3D flows over time get more extreme in the sense that a wider envelop of 3D flow ranges is observed as shown in Figures 7–9. Although an uncertainty does still exist in our assumption, the outcome from this research will provide a useful insight for water managers for their future water management practices based on scientific facts rather than personal judgement.

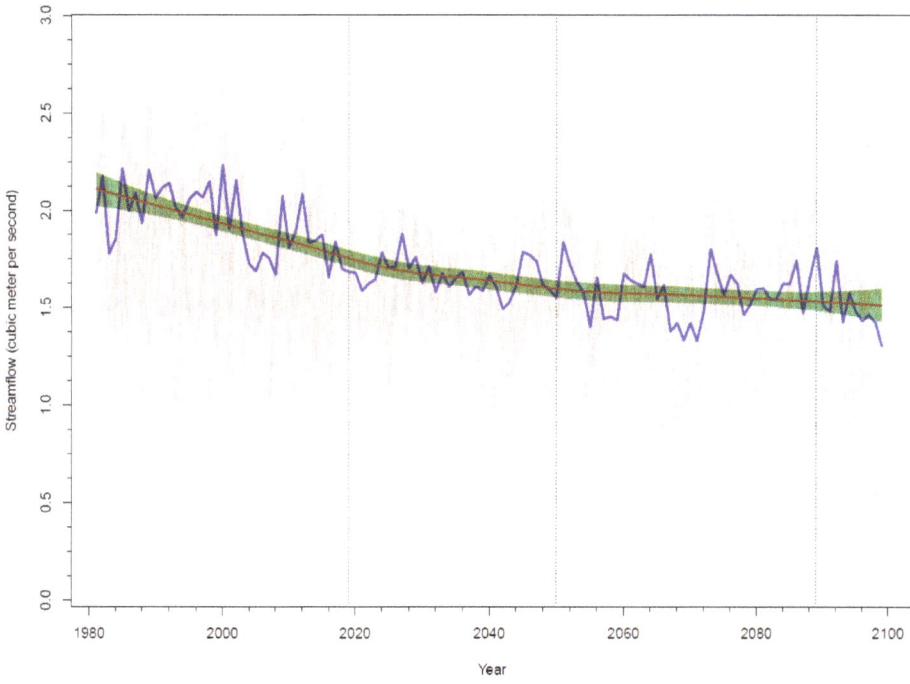

Figure 7. The climate-driven ensemble 3D flows at OBS1.

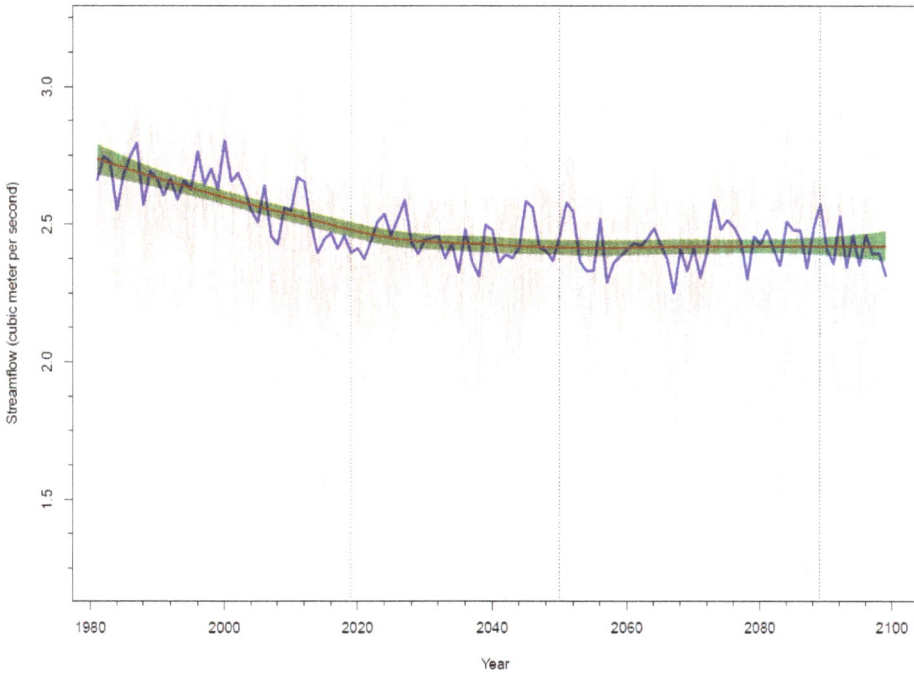

Figure 8. The climate-driven ensemble 3D flows at OBS2.

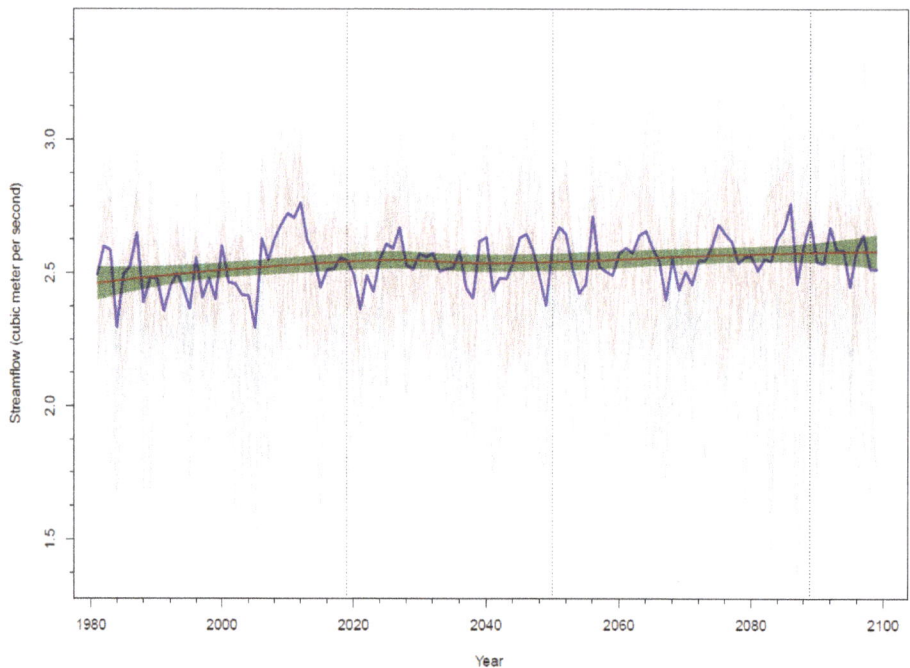

Figure 9. The climate-driven ensemble 3D flows at OBS3.

5. Conclusions

We have conducted a study on climate-driven flood risks in the Boise River Watershed using flood frequency analysis and future streamflow ensembles generated by HSPF with climate inputs. Three distribution families, including the Gumbel Extreme Value Type I (GEV), the 3-parameter log-normal (LN3) and log-Pearson type III (LP3) are used to predict future flood risks using a 3-day running total flow (3D flow). In addition to this conventional flood frequency analysis, climate-driven streamflow ensembles are also generated to oversee the likelihood of future flash flood events over the next few decades until 2099. The result indicates that the magnitude of the potential flash flood events is likely increasing over time from both methods, while climate-induced future ensemble streamflows (3D flows) is a broader envelop of historic flow variability. This implies that optimal use of available climate information should be practiced for water managers to plan out their adaptation strategies associated with hydroclimatic nonstationary and uncertainty in a changing global environment. We anticipate that this research will provide useful insights for water stakeholders to make a better decision based on scientific facts rather than personal conjecture. Furthermore, this study can be exemplified to explore future water storage design and management practices in the Boise River Watershed to cope with climate uncertainties.

Author Contributions: J.K. applied HSPF model to generate climate-induced hydrographs and J.H.R. proposed the study and contributed to conceptualizing the project, interpreting the processes in general as J.K.'s advisor.

Funding: This research is supported partially by the National Institute of Food and Agriculture, U.S. Department of Agriculture (USDA), under ID01507 and the Idaho State Board of Education (ISBOE) through IGEM program. Any opinions, findings, conclusions, or recommendations expressed in this publication are those of the authors and do not necessarily reflect the view of USDA and ISBOE.

Conflicts of Interest: The authors declare no conflict of interest.

Appendix A

$$R = \cfrac{\frac{1}{N} \times \sum_{i=1}^{N}\left(Q_{Oi} - \overline{Q}_{Oi}\right) \times \left(Q_{Si} - \overline{Q}_{Si}\right)}{\sqrt{\cfrac{N \times \sum_{i=1}^{N} Q_{Oi}^2 - \left(\sum_{i=1}^{N} Q_{O1}\right)^2}{N \times (N-1)}} \times \sqrt{\cfrac{N \times \sum_{i=1}^{N} Q_{Si}^2 - \left(\sum_{i=1}^{N} Q_{S1}\right)^2}{N \times (N-1)}}}, \tag{A1}$$

$$\text{NSE} = 1.0 - \left[\cfrac{\sum_{i=1}^{N}\left(Q_{Oi} - Q_{Si}\right)^2}{\sum_{i=1}^{N}\left(Q_{Oi} - \overline{Q}_{Oi}\right)^2}\right], \tag{A2}$$

$$\text{RSR} = \cfrac{\sqrt{\sum_{i=1}^{N}\left(Q_{Oi} - Q_{Si}\right)^2}}{\sqrt{\sum_{i=1}^{N}\left(Q_{Oi} - \overline{Q}_{Oi}\right)^2}}, \tag{A3}$$

$$\text{PBIAS} = \cfrac{\sum_{i=1}^{N}\left(Q_{Oi} - Q_{Si}\right)}{\sum_{i=1}^{N} QY_{Oi}} \times 100, \tag{A4}$$

where, Q_{Oi} and Q_{Si} are observed and simulated streamflow at the time step, respectively. $\overline{Q}_{Oi}$ and $\overline{Q}_{Si}$ are mean observed and simulated streamflow for the simulation period. N is the total number of values within the simulation period. R is the correlation coefficient between the predicted and observed values. It ranges from 0.0 to 1.0. A higher value indicates better agreement between predicted and observed data. Santhi et al. [58] indicated that R values greater than 0.7 show acceptable model performance. NSE is the percentage of the observed variance and determines the efficiency criterion for model verification [59]. It is calculated from minus infinity to 1.0. Higher positive values indicate better agreement between observed and simulated values. RSR is a standardized Root Mean Square Error (RMSE) based on observed standard deviation recommended by Legates and McCabe [60]. A zero value shows the optimal model performance. PBIAS calculates the average tendency of the simulated values to be larger or smaller than observed counterparts [61]. Lower PBIAS value (e.g., close to zero) indicates better performance. Positive PBIAS indicates underestimated bias, while negative PBIASO values shows the overestimated bias.

References

1. Das, T.; Maurer, E.P.; Pierce, D.W.; Dettinger, M.D.; Cayan, D.R. Increasing in flood magnitudes in California under warming climates. *J. Hydrol.* **2013**, *501*, 101–110. [CrossRef]
2. Elsner, M.M.; Cuo, L.; Voisin, N.; Deems, J.S.; Hamlet, A.F.; Vano, J.A.; Mickelson, K.E.B.; Lee, S.Y.; Lettenmaier, D.P. Implications of 21st century climate change for the hydrology of Washington State. *Clim. Chang.* **2010**, *102*, 225–260. [CrossRef]
3. Wahl, T.; Jain, S.; Bender, J.; Meyers, S.D.; Luther, M.E. Increasing risk of compound flooding from strom surge and rainfall for major US cities. *Nat. Clim. Chang.* **2015**, *5*, 1093–1097. [CrossRef]
4. Wing, O.E.J.; Bates, P.D.; Smith, A.M.; Sampson, C.C.; Johnson, K.A.; Fargione, J.; Morefield, P. Estimates of present and future flood risk in the conterminous United States. *Environ. Res. Lett.* **2018**, *13*, 034023. [CrossRef]
5. NOAA. Billion-Dollar Weather and Climate Disasters: Table of Events. 2017. Available online: https://www.ncdc.noaa.gov/billions/events/US/2017 (accessed on 28 January 2019).
6. Raff, D.A.; Pruitt, T.; Brekke, L.D. A framework for assessing flood frequency based on climate projection information. *Hydrol. Earth Syst. Sci.* **2009**, *13*, 2119–2136. [CrossRef]
7. Clow, D.W. Changes in the Timing of Snowmelt and Streamflow in Colorado: A Response to Recent Warming. *J. Clim.* **2010**, *23*, 2293–2306. [CrossRef]
8. Safeeq, M.; Grant, G.E.; Lewis, S.L.; Tague, C.L. Coupling snowpack and groundwater dynamics to interpret historical streamflow trends in the western United States. *Hydrol. Process.* **2012**, *27*, 655–668. [CrossRef]
9. NWS (National Weather Service). *Advanced Hydrologic Prediction Service* 2019. Available online: https://water.weather.gov/ahps2/inundation/index.php?gage=bigi1 (accessed on 1 March 2019).

10. Melillo, J.M.; Richmond, T.C.; Yohe, G.W. Climate Change Impacts in the United States: The third national climate assessment. U.S. *Glob. Chang. Res. Program* **2014**, *841*. [CrossRef]

11. Zhou, Q.; Leng, G.; Peng, J. Recent changes in the occurrences and damages of floods and droughts in the United States. *Water* **2018**, *10*, 1109. [CrossRef]

12. Dettinger, M.; Udall, B.; Georgakakos, A. Western water and climate change. *Ecol. Appl.* **2015**, *24*, 2069–2093. [CrossRef]

13. Kim, J.J.; Ryu, J.H. Modeling hydrological and environmental consequences of climate change and urbanization in the Boise River Watershed, Idaho. *J. Am. Water Resour. Assoc.* **2019**, *55*, 133–153. [CrossRef]

14. Pierce, D.W.; Cayan, D.R.; Das, T.; Maurer, E.P.; Miller, N.; Bao, Y.; Kanamitsu, M.; Yoshimura, K.; Snyder, M.A.; Sloan, L.C.; et al. The Key Role of Heavey Precipitation Events in Climate Model disagrremnts of Future Annual Precipitation Changes in California. *Am. Meteorol. Soc.* **2013**, *26*, 5879–5896.

15. USBR. Chapter 2: Hydrology and Climate Assessment. In *SECURE Water Act Section 9503(c)—Reclamation Climate Change and Water*; United States Bureau of Reclamation: Denver, CO, USA, 2016.

16. Weil, W.; Jia, F.; Chen, L.; Zhang, H.; Yu, Y. Effects of surficial condition and rainfall intensity on runoff in loess hilly area, China. *J. Hydorl.* **2014**, *56*, 115–126.

17. Miller, M.L.; Palmer, R.N. *Developing an Extended Streamflow Forecast for the Pacific Northwest*; World Water & Environmental Resources Congress ASCE: Philadelphia, PA, USA, 2003.

18. Wood, A.W.; Maurer, E.P.; Kumar, A.; Lettenmaier, D.P. Long-Range Experimental Hydrologic Forecasting for the Eastern United States. *J. Geophys. Res.* **2002**, *107*, ACL 6-1–ACL 6-15. [CrossRef]

19. Ryu, J.H. Application of HSPF to the Distributed Model Intercomparison Project: Case Study. *J. Hydrol. Eng.* **2009**, *14*, 847–857. [CrossRef]

20. Quintero, F.; Mantilla, R.; Anderson, C.; Claman, D.; Krajewski, W. Assessment of Changes in Flood Frequency Due to the Effects of Climate Change: Implications for Engineering Design. *Hydrology* **2018**, *5*, 19. [CrossRef]

21. IDEQ. *Lower Boise River Nutrient Subbasin Assessment*; Idaho Department of Environmental Quality: Boise, ID, USA, 2001; pp. 18–19.

22. Carlson, B. Proposal to Raise One of Three Boise River Dams Favors Anderson Ranch. Available online: https://www.capitalpress.com/state/idaho/proposal-to-raise-one-of-three-boise-river-dams-favors/article_52d1f19f-fbed-5843-9736-7cb3174b1573.html (accessed on 15 March 2019).

23. Tasker, G.D. A Comparison of methods for estimating low flow characterisics of streams. *J. Am. Water Resour. Assoc.* **1987**, *23*, 1077–1083. [CrossRef]

24. Loganathan, G.V. Frequency analysis of low flows. *Nord. Hydrol.* **1985**, *17*, 129–150. [CrossRef]

25. Matalas, N.C. Probability distribution of low flows, Statistical Studies in Hydrology. In *Geological Survey Professional Paper 434-A*; 1963. Available online: https://pubs.usgs.gov/pp/0434a/report.pdf (accessed on 1 March 2019).

26. Ryu, J.H.; Lee, J.H.; Jeong, S.; Park, S.K.; Han, K. The impacts of climate change on local hydrology and low flow frequency in the Geum River Basin, Korea. *Hydrol. Process.* **2011**, *25*, 3437–3447. [CrossRef]

27. Linsley, R.K.; Kohler, M.A.; Paulhus, J.L.H. *Hydrology for Engineers*; McGraw-Hill: New York, NY, USA, 1982.

28. Stedinger, J.R.; Vogel, R.M.; Efi, F.-G. *Handbook of Hydrology*; Maidment, D., Ed.; McGraw-Hill: New York, NY, USA, 1993; pp. 18.12–18.13.

29. Weibull, W.A. *Statistical Theory of the Strength of Materials*; the Royal Swedish Institute for Engineering Research: Stockholm, Swedish, 1939.

30. Chow, V.T. A general formula for hydrologic frequency analysis. *Trans. Am. Geophys. Union* **1951**, *32*, 231–237. [CrossRef]

31. Fisher, R.A.; Tippett, L.H.C. Limiting Forms of the Frequency Distributions of the Smalles and Largest Member of a Sample. *Proc. Camb. Philos. Soc.* **1928**, *24*, 180–190. [CrossRef]

32. Hosking, J.R.M.; Wallis, J.R.; Wood, A.W. Estimation of the Generalized Extreme Value Distribution by the Method of Probability-Weighted Moments. *Technometrics* **1985**, *27*, 251–261. [CrossRef]

33. Hoshi, K.; Stedinger, J.R.; Burges, S.J. Estimation of Log-Normal Quantiles: Monte Carlo Results and First-Order Approximations. *J. Hydrol.* **1984**, *71*, 1–30. [CrossRef]

34. Stedinger, J.R. Fitting Log Normal Distributions to Hydrologic Data. *Water Resour. Res.* **1980**, *16*, 481–490. [CrossRef]

35. Bobee, B. The Log Pearson Type 3 Distribution and Its Application in Hydrology. *Water Resour. Res.* **1975**, *11*, 681–689. [CrossRef]
36. Dudula, J.; Randhir, T.O. Modeling the influence of climate change on watershed systems: Adaptation through targeted practices. *J. Hydrol.* **2016**, *541*, 703–713. [CrossRef]
37. Tong, S.T.Y.; Ranatunga, T.; He, J.; Yang, T.J. Predicting plausible impacts of sets of climate and land use change scenarios on water resources. *Appl. Geogr.* **2012**, *32*, 477–489. [CrossRef]
38. Stern, M.; Flint, L.; Minear, J.; Flint, A.; Wright, S. Characterizing Changes in Streamflow and Sediment Supply in the Sacramento River Basin, California, Using Hydrological Simulation Program-Fortran (HSPF). *Water* **2016**, *8*, 432. [CrossRef]
39. Bicknell, B.; Imhoff, J.; Kittle, J., Jr.; Jones, T.; Donigian, A., Jr.; Johanson, R. *Hydrological Simulation Program-FORTRAN: HSPF Version 12 User's Manual*; AQUA TERRA Consultants: Mountain View, CA, USA, 2001.
40. Crawford, N.H.; Linsley, R.K. *Digital Simulation in Hydrology: Stanford Watershed Model IV*; Technical Report No. 39; Department of Civil Engineering, Stanford University: Stanford, CA, USA, 1966.
41. Donigian, A.S.; Davis, H.H. *User's Manual for Agricultural Runoff Management (ARM) Model*; The National Technical Information Service: Springfield, VA, USA, 1978.
42. Donigian, A.S.; Crawford, N.H. *Modeling Nonpoint Pollution from the Land Surface*; US Environmental Protection Agency, Office of Research and Development, Environmental Research Laboratory: Washington, DC, USA, 1976.
43. Donigian, A.S.; Huber, W.C.; Laboratory, E.R.; Consultants, A.T. *Modeling of Nonpoint Source Water Quality in Urban and Non-urban Areas*; US Environmental Protection Agency, Office of Research and Development, Environmental Research Laboratory: Washington, DC, USA, 1991.
44. Donigian, A., Jr.; Bicknell, B.; Imhoff, J.; Singh, V. Hydrological Simulation Program-Frotran (HSPF). In *Computer Models of Watershed Hydrology*; Water Resources Publications: Highlands Ranch, CO, USA, 1995; pp. 395–442.
45. Kim, J.J.; Ryu, J.H. A threshold of basin discretization for HSPF simulations with NEXRAD inputs. *J. Hydrol. Eng.* **2013**, *19*, 1401–1412. [CrossRef]
46. Mitchell, K.E.; Lohmann, D.; Houser, P.R.; Wood, E.F.; Schaake, J.C.; Robock, A.; Cosgrove, B.A.; Sheffield, J.; Duan, Q.Y.; Luo, L.; et al. The multi-institution North American Land Data Assimilation System (NLDAS): Utilizaing multiple GCIP products and partners in a continental distributed hydrologicl modeling system. *J. Geophys. Res.* **2004**, *109*, D07S90. [CrossRef]
47. Abatzoglou, J.T. Development of gridded surface meteorological data for ecological applications and modeling. *Int. J. Climatol.* **2011**, *33*, 121–131. [CrossRef]
48. Fasano, G.; Franceschini, A. A multivariate version of the Kolmogorov-Sminov test. *Mon. Not. R. Astron. Soc.* **1987**, *225*, 155–170. [CrossRef]
49. Benjamin, J.R.; Cornell, C.A. *Probability, Statistics and Decision for Civil Engineers*; McGraw-Hill: New York, NY, USA, 1970; 684p.
50. Kite, G.W. Confidence limits for design events. *Water Resour. Res.* **1975**, *11*, 48–53. [CrossRef]
51. Duda, P.B.; Hummel, P.R.; Donigian, A.S., Jr.; Imhoff, J.C. BASINS/HSPF: Model use, calibration, and validation. *Trans. ASABE* **2012**, *55*, 1523–1547. [CrossRef]
52. Moriasi, D.N.; Arnold, J.G.; Van Liew, M.W.; Binger, R.L.; Harmel, R.D.; Veith, T.L. Model Evaluation Guidelines for Systematic Quantification of Accuracy in Watershed Simulations. *Am. Soc. Agric. Biol. Eng.* **2007**, *50*, 885–900. [CrossRef]
53. Milly, P.C.D.; Betancourt, J.; Falkenmark, M.; Hirsch, R.M.; Kundzewicz, Z.W.; Letternmaier, D.P. Stationarity Is Dead: Whiher Water Management? *Science* **2008**, *319*, 573–574. [CrossRef]
54. Veldkamp, T.; Frieler, K.; Schewe, J.; Ostberg, S.; Willner, S.; Schauberger, B.; Gosling, S.N.; Schmied, H.M.; Portmann, F.T.; Huang, M.; et al. The critical role of the routing scheme in simulating peak river discharge in global hydrological models. *Environ. Res. Lett.* **2017**, *12*, 075003.
55. Liu, S.; Huang, S.; Xie, Y.; Wang, H.; Leng, G.; Huang, Q.; Wei, X.; Wang, L. Identification of the non-stationarity of floods: Changing patterns, causes, and implications. *Water Resour. Manag.* **2019**, *33*, 939–953. [CrossRef]
56. Fcinocca, J.F.; Kharin, V.V.; Jiao, Y.; Qian, M.W.; Lazare, M.; Solheim, L.; Flato, G.M.; Biner, S.; Desgagne, M.; Dugas, B. Coordinated global and regional climate modeling. *J. Clim.* **2016**, *29*, 17–35. [CrossRef]

57. Kim, J.J.; Ryu, J.H. Quantifying the performances of the semi-distributed hydrologic model in parallel computing—A case study. *Water* **2019**, *11*, 823. [CrossRef]
58. Santhi, C.; Arnold, J.G.; Williams, J.R.; Dugas, W.A.; Srinivasan, R.; Hauck, L.M. Validation of the SWAT Model on a Large River Basin with Point and Nonpoint Sources. *J. Am. Water Resour. Assoc.* **2001**, *37*, 1169–1188. [CrossRef]
59. Nash, J.E.; Sutcliffe, J.V. River flow forecasting through conceptual models: Part I—A discussion of principles. *J. Hydrol.* **1970**, *10*, 282–290. [CrossRef]
60. Legates, D.R.; McCabe, G.J. Evaluating the Use of 'Goodness-of-Fit' Measures in Hydrologic and Hydroclimatic Model Validation. *Water Resour. Res.* **1999**, *35*, 233–241. [CrossRef]
61. Gupta, I.; Gupta, A.; Khanna, P. Genetic algorithm for optimization of water distribution system. *Environ. Model. Softw.* **1999**, *14*, 437–446. [CrossRef]

Article

Assessment of Climate Change Impacts on Extreme High and Low Flows: An Improved Bottom-Up Approach

Abdullah Alodah [1,2,*] and Ousmane Seidou [1]

[1] Department of Civil Engineering, Faculty of Engineering, University of Ottawa, 161 Louis Pasteur, Ottawa, ON K1N 6N5, Canada; Ousmane.Seidou@uottawa.ca
[2] Department of Civil Engineering, Faculty of Engineering, Qassim University, Buraydah, 51431 Al Qassim, Saudi Arabia
* Correspondence: asalodah@gmail.com

Received: 2 May 2019; Accepted: 10 June 2019; Published: 13 June 2019

Abstract: A quantitative assessment of the likelihood of all possible future states is lacking in both the traditional top-down and the alternative bottom-up approaches to the assessment of climate change impacts. The issue is tackled herein by generating a large number of representative climate projections using weather generators calibrated with the outputs of regional climate models. A case study was performed on the South Nation River Watershed located in Eastern Ontario, Canada, using climate projections generated by four climate models and forced with medium- to high-emission scenarios (RCP4.5 and RCP8.5) for the future 30-year period (2071–2100). These raw projections were corrected using two downscaling techniques. Large ensembles of future series were created by perturbing downscaled data with a stochastic weather generator, then used as inputs to a hydrological model that was calibrated using observed data. Risk indices calculated with the simulated streamflow data were converted into probability distributions using Kernel Density Estimations. The results are dimensional joint probability distributions of risk-relevant indices that provide estimates of the likelihood of unwanted events under a given watershed configuration and management policy. The proposed approach offers a more complete vision of the impacts of climate change and opens the door to a more objective assessment of adaptation strategies.

Keywords: hydrological risk assessment; extreme hydrologic events; climate change impacts; downscaling; uncertainty; ensembles; water resource systems

---

1. Introduction

Climate change has forced the research community to revisit assumptions and theories used for the design, planning, and management of water resource systems [1]. Hydrological processes depend to a great degree on the climate regime [2–7]. Perturbing the climate regime inevitably results in a perturbed hydrological regime, which in turn affects water-related risk and water resource system performance. Numerous researchers and practitioners have been working intensely to develop methodologies to identify future climate change impacts and viable adaptations strategies [8–10].

The most frequently used approach in climate impact assessments is the top-down approach, which is constrained by the availability of Global Circulation Models (GCM) and Regional Climate Models (RCM). Typically, it considers possible future changes in climatic conditions based on predetermined scenarios that are parameterized in large-scale models. This is achieved by downscaling a few distinct projections to be able to determine the impacts on the local hydrological system. However, a critical limitation of this approach is that it ignores plausible risks by not covering all possible future conditions despite using multi-model and multi-scenario projections. For example, while climate projections are mainly derived from multidecadal GCM simulations, the latter poorly account for many natural climate forcings, such as volcanic eruptions, which lead to added uncertainty, particularly when estimating local and regional impacts [11]. Also, all downscaled scenarios are equally likely, which makes it impossible to choose one scenario over another.

The alternative bottom-up approach uses a wide range of possible conditions to assess the sensitivity of water resource systems to climate change and identify potentially risky situations [12]. This approach was only recently introduced to the field of climate adaptations in water research, where it has been gaining traction, particularly in risk assessment and planning [13–18]. However, the bottom-up approach creates large and uncertain vulnerability domains that are inconvenient for the decision-maker. In addition, both approaches suffer from the limited availability of GCM/RCM projections at a given watershed, and there is no existing methodology to rank their outputs [19]. These limitations are real hindrances when considering such projections in a decision or design framework and need to be improved.

By using modelled climate data, precipitation-runoff models are frequently used to simulate hydrological processes, quantify potential impacts of climate change, and identify water availability issues, particularly under extreme conditions. There are two common types of extreme streamflow that are usually considered in impact studies: peak events (causing flooding) using frequency analysis and low streamflow indices (causing drought) [20,21]. Extreme high and low flow indicators based on return periods are used for efficient and safe engineering designs. The importance of an engineering structure and the consequences of its potential failure determine the choice of the return period (e.g., dams are designed based on a 1000-year return period). Besides the use of low-flow information in typical water sector applications such as energy, irrigation, and navigation, there are also noticeably increasing efforts to study risks imposed upon an aquatic habitat and to regulate minimum environmental flow requirements and sustain water quantity and quality [22–25]. Also, low-flow indices are used to prevent deterioration of freshwater ecosystems, which are very vulnerable to climate change.

In order to address the aforementioned unresolved issues, this paper proposes a methodology to generate a large number of climate change projections by combining the outputs of stochastic weather generators and regional climate models to create an ensemble of projections with quantifiable probabilities. This new set of projections provides better coverage of the risk space and can facilitate the implementation of a bottom-up approach by considering the plausibility of risk, which may lead to informed decisions and robust adaptation plans.

2. Materials and Methods

One of the major limitations of the top-down approach is that the number of scenarios is too low to fully explore the climate-related risk space. The problem can be partially mitigated by cloning projections generated by corrected climate models and by slightly perturbing them through a stochastic weather generator to obtain a larger set of scenarios that cover the same statistical space as the synthetic time series.

Here is a brief explanation of the 'ensemble generation process' used herein for risk discovery. We firstly used downscaling methods to correct biased (i.e., raw) regional climate model data. Secondly, using corrected (i.e., downscaled) climate data, the stochastic weather generator (MulGETS) was utilized to generate 30-years of daily precipitation and minimum and maximum temperature data. The second step is repeated to generate a large number of climate projections: we run MulGETS 250 with each corrected RCM data. Thirdly, the calibrated SWAT model was forced with these climate projections individually to generate "ensemble" of streamflow projections. Subsequently, the newly-generated ensemble was utilized to determine the risk space that is largely overlooked by top-down methods. This was done to consider the inherent randomness of hydroclimatic variables. Figure 1 presents a schematic representation of all datasets used in this study and the relationships between different time series in the proposed framework. The subsequent sections describe the aforementioned process in details.

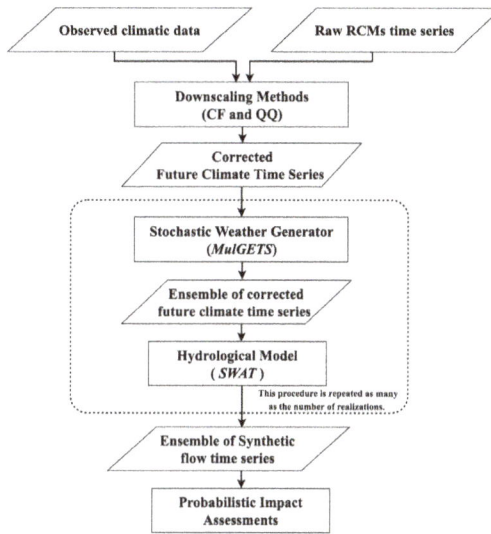

Figure 1. Schematic representation of the proposed approach.

2.1. Study Area

The South Nation Watershed (SNW) of about 4000 km^2 in Eastern Ontario was used as a showcase. It is characterized by multigauge climatic data (from St. Albert, Russell, Morrisburg, and South Mountains metrological stations) and downstream daily discharge data (collected at Plantagenet gauging station (ID: 02LB005)) (Figure 2). Hydrogeological investigations have indicated that an impermeable overburden watershed and a lack of streambed gradients in parts of the South Nation River pose a prominent flooding risk [26]. The average maximum and minimum air temperatures are 11.5 and 1.2 °C, respectively, with a mean annual precipitation of around 1000 mm. A more detailed description of the watershed has been presented previously [27].

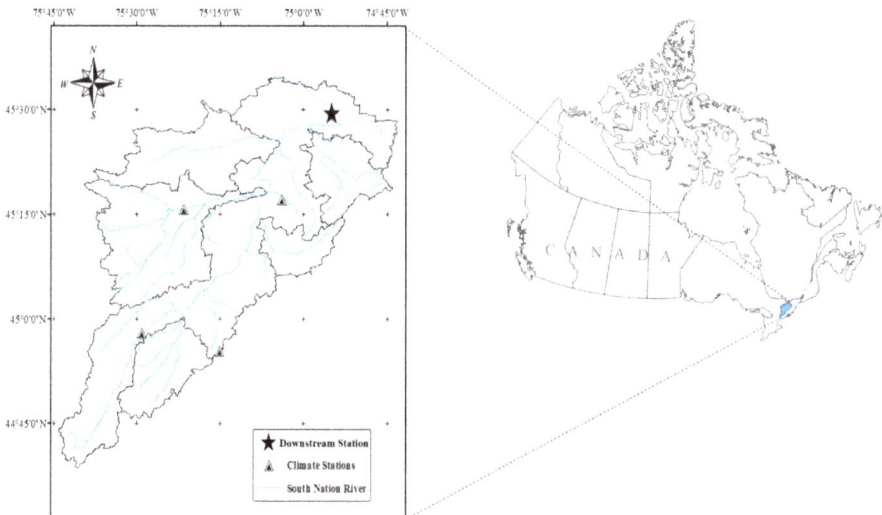

Figure 2. Map of the South Nation Watershed.

2.2. Regional Climate Model Data

The availability of daily climatic variables at fine spatial scales is essential for hydrological impact studies [28]. The gridded daily data of precipitation, maximum near-surface air temperature, and minimum near-surface air temperature were extracted from the North American domain of Coordinated Regional Climate Downscaling Experiment (NA-CORDEX) project archive, driven by state-of-the-art GCMs as part of the the Coupled Model Inter-comparison Project Phase 5 (CMIP5) experiment (Table 1). The CORDEX project was initiated by the World Climate Research Program (WCRP) to provide reliable predictions of regional future climate change. The four independent models were forced with medium- to high-emission scenarios RCP4.5 and RCP8.5 (resulting in 8 different projections). They are originally based on outputs from the CanESM2 and EC-EARTH Global Circulation Models, whose outputs are downscaled with three Regional Climate Models: CanRCM4, RCA4, and HIRHAM5. The horizontal spatial resolution was chosen to be NAM-44 (North American domain at 0.44° horizontal grid resolution or approximately 50 km).

Table 1. List of Regional Climate Models used in the study.

Global Circulation Models (GCM) (Driver)	Regional Climate Model (RCM)	Grid (Spatial Resolution)	Representative Concentration Pathways	Variables
Second-generation Canadian Earth System Model (CanESM2)	Canadian Regional Climate Model (CanRCM4)	NAM-44 (0.44°)	RCP4.5 RCP8.5	pr; tasmax; tasmin
Second-generation Canadian Earth System Model (CanESM2)	Rossby Centre Regional Atmospheric Climate Model (RCA4)	NAM-44 (0.44°)	RCP4.5 RCP8.5	pr; tasmax; tasmin
European Earth System Model (EC-EARTH)	Danish Climate Centre Model (HIRHAM5)	NAM-44 (0.44°)	RCP4.5 RCP8.5	pr; tasmax; tasmin
European Earth System Model (EC-EARTH)	Rossby Centre Regional Atmospheric Climate Model (RCA4)	NAM-44 (0.44°)	RCP4.5 RCP8.5	pr; tasmax; tasmin

2.3. Downscaling Methods

Downscaling is a procedure undertaken to reconcile the large-scale data representativeness of observations and produce climate change data at a finer resolution. This study makes use of two statistical downscaling techniques—namely Change Factors (CF) and Quantile-Quantile transformation (QQ)—to produce more reliable station-level climatic information to be used later as inputs to a distributed impact model.

The linear correction method applied in this study, namely CF, requires three time series: observations (or synthetic climate series), historical, and future raw RCM data. The primary assumption in this method is that the relationship between historical and future projections is the same as the relation between historical RCM data and observed time series. However, the Quantile-Quantile method applies a different concept that is based on the distribution of raw simulations and observations. For a given variable, downscaling is conducted using a monthly time-step. Observed precipitation and temperature variables of 30 years (1981 to 2010) were used to define a reference climatology and to spatially downscale the coarse-scale climate data in the future (2071–2100).

2.3.1. Change Factor Method

The Change Factors (CF) method is a commonly used approach to linearly downscale a variable (e.g., precipitation) by calculating the difference between the control and future GCM simulations before scaling the baseline observations accordingly [29]. Yet, the conventional CF method is usually used a top-down approach and as such not always reliable. Nevertheless, it has been used excessively in recent climate change assessments [30–37]. The CF approach lacks the ability to correct future projections as it predicts the same variability for the future climate by keeping the same historical temporal structure, which is unrealistic [29,30,38]. One of its limitations is that it produces the exact same temporal structure of wet days (occurrences) in the future [39], which is by no means certain for a stochastic variable such as precipitation. Furthermore, extreme events that are vital in risk assessments cannot be adequately corrected by such a method when using only one observational dataset. To alleviate these disadvantages, we employed stochastic weather generators as they are capable of capturing climate variability. We explored the potential risks by producing a new set of weather data and applying factors of change obtained from raw daily RCM outputs between the reference period and the future.

Each climate station is perturbed using the simplest version of the CF method, that is, we applied the following deterministic transformation to downscaled climate model outputs:

$$PCP_t^{RCM, CF} = PCP_t^{OBS} \frac{AVG_{historical}\left(PCP_t^{RCM}\right)}{AVG_{future}\left(PCP_t^{RCM}\right)}, \tag{1}$$

where $PCP_t^{RCM, CF}$ is the perturbed RCM output, PCP_t^{RCM} is the original RCM output, and PCP_t^{OBS} denotes the observed data. Analogously for temperature, we use:

$$TMP_t^{RCM, CF} = TMP_t^{OBS} + \left[AVG_{future}\left(TMP_t^{RCM}\right) - AVG_{historical}\left(TMP_t^{RCM}\right)\right], \tag{2}$$

where $TMP_t^{RCM, CF}$ is the perturbed RCM output, TMP_t^{RCM} is the original RCM output, and TMP_t^{OBS} is the observed climate. The expectation is that after transformation, the means of $PCP_t^{RCM, CF}$ and $TMP_t^{RCM, CF}$ will each be closer to the respective means of PCP_t^{OBS} and TMP_t^{OBS} over the reference period.

2.3.2. Quantile-Quantile Transformation

Quantile-Quantile (QQ) transformation is an empirical routine applied to correct systematic errors in regional climate models [33,40,41]. The raw RCM outputs typically do not have the same distribution as the observations [42], and both distributions must be reconciled. It is more sophisticated than mean-based methods, and its procedure includes remapping the probability density function (PDF) of uncorrected RCM data onto the PDF of observations. Corrected RCM simulations, X_{CORR}, of the future are produced by applying the following transformation to their cumulative distribution functions (F):

$$X_{CORR} = F_{OBS}^{-1} \left(F_{RCM}\left(X_{RCM}\right)\right), \tag{3}$$

where X_{RCM} refers to the climate variable extracted from raw RCM data. This technique has been proven to be more accurate for reproducing a valid agreement between the corrected and observed PDFs than the linear method discussed earlier [29,41–43]. However, one of the main drawbacks of the QQ technique is its inability to predict values beyond the original range of historic extremes [44], which may ultimately affect risk analysis.

2.4. Generating an Ensemble of Corrected-RCM-Like Realizations

The Change Factors and Quantile-Quantile downscaling methods lack the ability to produce more than one possible future state and underestimates the impact variability [45]. The variability issue is tackled in this paper by generating an ensemble of corrected RCM-like realizations, as illustrated in Figure 1, which consists of multiple runs (the number of realizations is 250) of a given stochastic weather generator fed with corrected future climate for a 30-year period. This produces a total of 2000 unique future climate projections (= 250 × 4 × 2) for each downscaling method (= 2000 × 2). Methods involving stochastic weather generators are generally capable of creating a limitless number of sequences of weather data with novel scenarios which helps with the uncertainty analysis [31].

The selection of an adequate stochastic weather generator is crucial to efficiently explore a wide range of climate risk scenarios for water resource systems [21].

This study made use of the Multi-site weather Generator of the École de Technologie Supérieure (MulGETS) [46], a multisite, multivariate weather generator that correctly reproduced historical climatic and hydrological characteristics for the SNW [27]. Wet-day precipitation sequences were reproduced using a multi-Gamma distribution (a combination of several gamma distributions). MulGETS can account for the coherency among multiple climate variables. Details of climatic data and MulGETS configuration have been previously presented in Alodah and Seidou [27]. The final result is a super-ensemble combining all ensembles containing a large number of scenarios (hereafter the "ensemble") that inherit the trends projected by climate models while covering the range of variability from the historical period. Given that the variability range of the historical period is matched, the assumption that all scenarios are equally probable will now be more defendable.

2.5. Hydrological Response

Ultimately, the development of such new future-climate scenarios, essentially driven by corrected GCM/RCM projections, enables us to discover the risks presented by changing hydrology using altered flows at a local scale. The Soil and Water Assessment Tool (SWAT-2012) was used to simulate hydrological variables and utilizing the semi-automated SUFI-2 optimization algorithm (Sequential Uncertainty Fitting ver. 2) to evaluate the most accurate simulation based on an uncertainty analysis routine [47]. SWAT is a semi-distributed, watershed-scale hydrological model that has been extensively utilized to address water quality and quantity issues [48]. Hydrologists, conservationists, and policy makers have extensively used SWAT to predict a variety of water-related issues and their environmental impacts [48–55]. Weather information is the main physically-based input that controls the transformation of precipitation into runoff in SWAT [47].

The SNW was partitioned into 31 different Hydrologic Response Units (HRUs), each with its own land, management, and soil characteristics. A detailed description of the SWAT model configuration and parameterization used in this study has been presented previously in Alodah and Seidou [27]. A set of different statistic measures was used to assess the goodness of fit of the calibrated model, including the Nash–Sutcliffe efficiency, the RMSE-observations standard deviation ratio, and the percent bias (Table 2). Hydrological impact models, distributed models in particular (e.g., SWAT), are particularly sensitive to small-scale climate variations that might be underestimated by large-scale models [56], and it could be reasonably argued that relying on a selected few models, either in a top-down or bottom-up approach, cannot be adequately justified.

Table 2. Goodness-of-Fit metrics used for SWAT model performance evaluation.

Statistic Measure	Formula [1]
The Nash–Sutcliffe efficiency coefficient (NSE)	$NSE = 1 - \dfrac{\sum_{i=1}^{n} (O_i - P_i)^2}{\sum_{i=1}^{n} (O_i - \overline{O})^2}$
The RMSE-observations standard deviation ratio (RSR)	$PSR = \dfrac{\sqrt{\sum_{i=1}^{n} (O_i - P_i)^2}}{\sqrt{\sum_{i=1}^{n} (O_i - \overline{O})^2}}$
The percentage of bias (PBIAS)	$PBIAS = \dfrac{\sum_{i=1}^{n} (O_i - P_i) \times 100}{\sum_{i=1}^{n} (O_i)}$

[1] where O_i stands for observed and P_i for predicted values; $\overline{O}$ is the mean of the observed values.

2.6. Quantifying the Risk Spaces

Because the goal is to build on the bottom-up approach, the same framework as Brown et al. [57] was adopted. A climate state represented by a time series is summarized by subsets of climate and hydrological statistics that are relevant to the problem under investigation. These subsets can be calculated for any observational time series or any downscaled climate model outputs. A climate state will yield a level of risk and performance that is measured by a set of risk/performance indicators.

These indicators are obtained by feeding SWAT with the perturbed time series data, each of which is unique. We placed a greater focus on hydrological risks, due to their direct impacts on the watershed systems. This includes analyzing extreme values based on statistical models and investigating the timing and intensity of peak spring flow.

2.6.1. Extreme Value Statistical Probability Models (AM and 7Q)

The 3-parameter Generalized Extreme Value (GEV) and 2-parameter Weibull (WBL) distribution functions were fitted to the annual maximum streamflows (AM) and minimum extreme events (the lowest 7-day average flow, or 7Q), respectively, based on seven time-intervals (2, 5, 10, 20, 50, 100, and 500 years). The cumulative distribution functions (CDFs) of the GEV and WBL models are [58]:

$$F_{GEV}(x) = \exp\left\{-\left[1 + \xi\left(\frac{x-\mu}{\sigma}\right)\right]^{-1/\xi}\right\}, \text{ where } \left[1 + \xi\left(\frac{x-\mu}{\sigma}\right)\right] > 0, \tag{4}$$

$$F_{WBL}(x) = 1 - \exp\left\{-\left[\frac{x}{\sigma}\right]^{\xi}\right\} \tag{5}$$

where ξ, μ, and σ are the shape, location, and scale parameters, respectively. Maximum likelihood (ML) estimators, as recommended by Das et al. [59], are used to compute the parameters of these statistical models based on 95% confidence intervals ($\alpha = 0.05$).

2.6.2. Spring Flow Timing and Intensity

Another important phenomenon with significant impacts on the hydrology of many river systems in temperate regions is the timing of the annual snowmelt and changes to river-ice conditions. In the SNW, the most common flooding events are caused by snowmelt-driven runoff. Hence, future changes in the spring snowmelt and river-ice breakup, particularly at high latitudes or in large mountain regions, seem inevitable. These changes in river-ice processes are useful indicators of climate change because of their sensitivity to air temperature [60,61]. Warmer winters may promote dramatic alterations in discharge patterns and in the severity and timing of the snowmelt. A threshold amount of energy, including the mean daily temperature and soil heat flux, as well as the areal coverage of snow are critical in controlling the release of the stored water. In the SWAT model, the snowmelt starts when the daily maximum temperature (SMTMP) is above 0 °C and its uniformly released water is considered to be precipitation [48]. For the sake of simplicity, the peak spring flow is defined as the event of maximum daily flow in the spring season from January 1 to May 31.

2.7. Likelihood Estimation Using KDE

It is assumed that the synthetically generated time series will be a fair representation of the natural climate variability. Using all sample data, we employed a nonparametric density estimation function, or the kernel density estimator (KDE) to build a continuous probability density function (PDF) without making any a priori assumptions with regard to the underlying distribution. This approach involves exploring the extent of possible changes to hydroclimatic variables in the context of climate change. This includes analyzing the main characteristics of hydrological variables. The kernel density estimator of a sample of a d-variate random vector $(x_1, x_2, \ldots, x_n)$, drawn from an unknown distribution, is given by:

$$\hat{f}_h(x) = \frac{1}{n} \sum_{i=1}^{n} K(x - x_i), \tag{6}$$

where K is a non-negative kernel smoothing function, and h the non-negative bandwidth. The bivariate kernel density estimation in this study was based on a diagonal bandwidth matrix and a Gaussian kernel [62].

3. Results

3.1. SWAT Calibration and Validation

The results of the proposed method for the South Nation Watershed under changing climate conditions using a stochastic data-oriented approach are presented. Applying the two downscaling methods described above to different CORDEX data with corrected daily time series of precipitation, maximum and minimum temperature from four RCMs under the conditions of RCP4.5 and RCP8.5 emission scenarios were generated for every metrological station. The SWAT model was calibrated with 15 years of observed streamflow at a monthly resolution for the period from Jan. 1981 to Dec. 1995 and validated on the period from Jan. 1996 to Dec. 2005. Table 3 provides results of a set of statistics that show a satisfactory agreement between observed and simulated streamflows in both periods.

Table 3. SWAT model performance in the calibration and validation periods.

Statistic Measure	Preferred Ranges *	Period	
		Calibration	Validation
The Nash–Sutcliffe efficiency coefficient (NSE)	NSE ≥ 0.75	0.90	0.81
The RMSE-observations standard deviation ratio (RSR)	RSR ≤ 0.5	0.31	0.43
The percentage of bias (PBIAS)	$-10\% \leq PBIAS \leq 10\%$	−10.0%	−8.3%

* According to Liew et al. [63] and Moriasi et al. [64].

3.2. Time Series Generation for the Reference and Future Periods

The weather generator, MulGETS, was used to generate 250 30-year climate datasets for each climate scenario. Then, the SWAT model, calibrated on historical climate data, was forced with the corrected future climate information to investigate hydrological changes. This resulted in 4000 realizations of future daily streamflow sequences, each consisting of 30 years of predictions. The exact number of simulations is presented in Table 4. To account for uncertainties, the total of 480,000 years of predictions are summarized using extreme-streamflow indicators that describe the associated risks and allow for comparisons against observations.

Table 4. Details of models and number of future predictions used in the study.

Downscaling Methods	Climate Model: GCM (RCM)	RCPs	Variables (Daily)	Weather Generator	Hydrological Model	Period Length	Total (yrs)
Change Factor (CF) Quantile-Quantile (QQ)	CanESM2 (CanRCM4) CanESM2 (RCA4) EC-EARTH (HIRHAM5) EC-EARTH (RCA4)	RCP4.5 RCP8.5	PCP TMAX TMIN Streamflow	MulGETS	SWAT	30 (yrs) (2071–2100)	
(Total) 2	4	2	4	250 *	1	30	= 480,000

* Number of realizations used.

3.3. Representation of the Sensitivity Space

3.3.1. Flood and Drought Indicators

Hydrologically-based extreme streamflow metrics simulated by SWAT, including the design flood and 7-day low flow based on selective return periods, were analyzed. These results were compared to the equivalent SWAT-estimated extreme indices on the historical period to quantify possible changes (Figure 3). The isolines in Figure 3 depict the intensity of projections as the probability scale is given by the colour varying smoothly from yellow (higher intensity) to blue (less intensity). The analysis of both datasets showed that flooding events will be affected more than low flow extremes, presumably due to predicted changes in the duration of snow cover. Results of the ensemble (Figure 3e) imply significant decreases in the annual maximum flows and increasing summer minimum flows by mostly all data configurations (i.e., RCMs, RCPs, and downscaling methods).

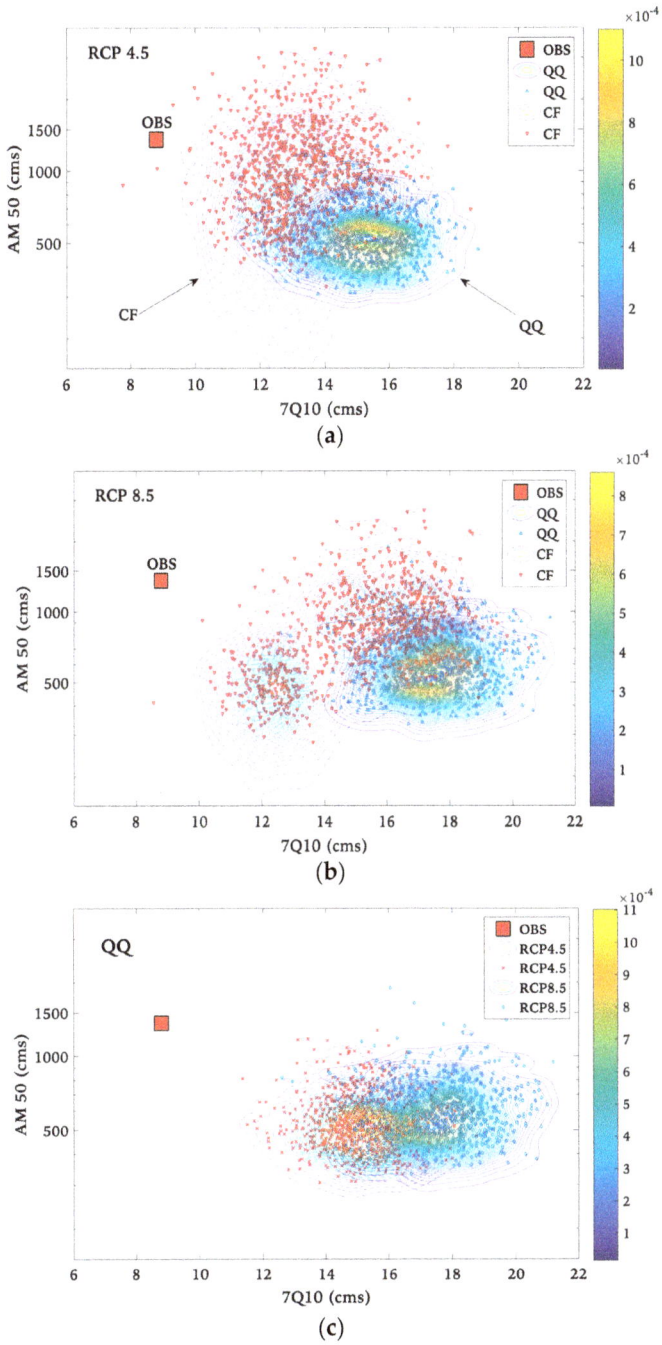

Figure 3. *Cont.*

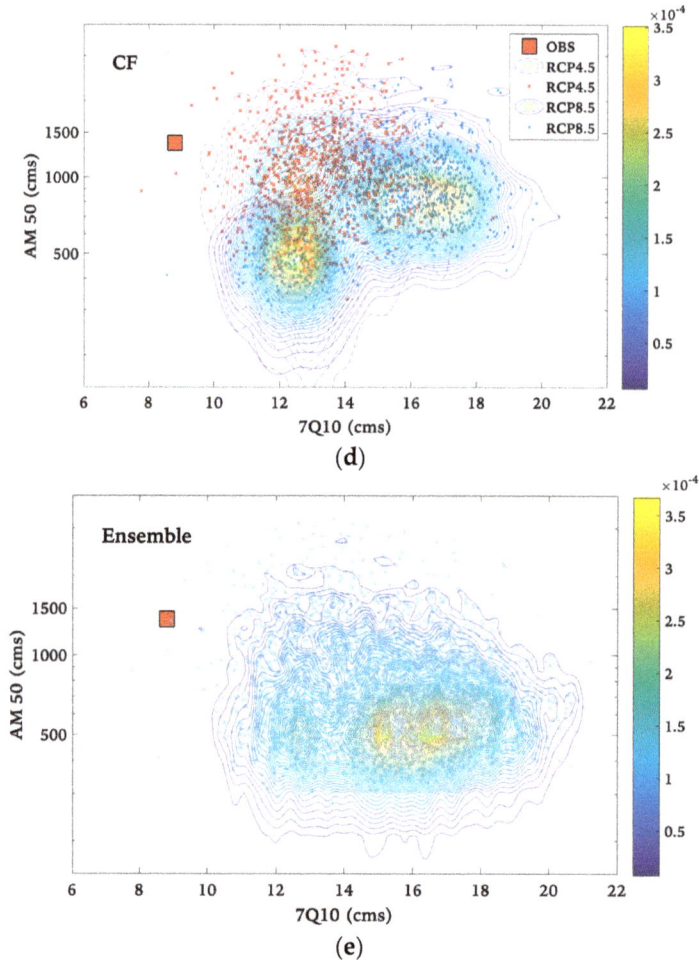

Figure 3. Bivariate kernel density estimations of the 50-year flooding event (AM 50) and the 7-day low flow of 10-year return period (7Q10) based on (**a**) RCP4.5 and (**b**) RCP8.5 scenarios, derived from ensembles of downscaled climate data using (**c**) QQ method, (**d**) CF method, and (**e**) the ensemble of all simulations. Each point represents a hydrological response to a climate-change realization (projection), and isolines represent their probability where isoline-values coded by color from yellow (higher intensity) to blue (less intensity). Results are compared to observed data (red square).

Further analysis was conducted on extreme values by combining all simulations by each downscaling method, resulting in 30,000 years of simulated streamflow data. Drought indices (7Q) and design floods (AM) of these time series were compared to observed ones (Figure 4). The comparison indicates that higher air temperatures lead to reduced summer low-flows and less intense high flows. The ensemble results indicate similar trends in both indices—fewer droughts and extreme flooding events.

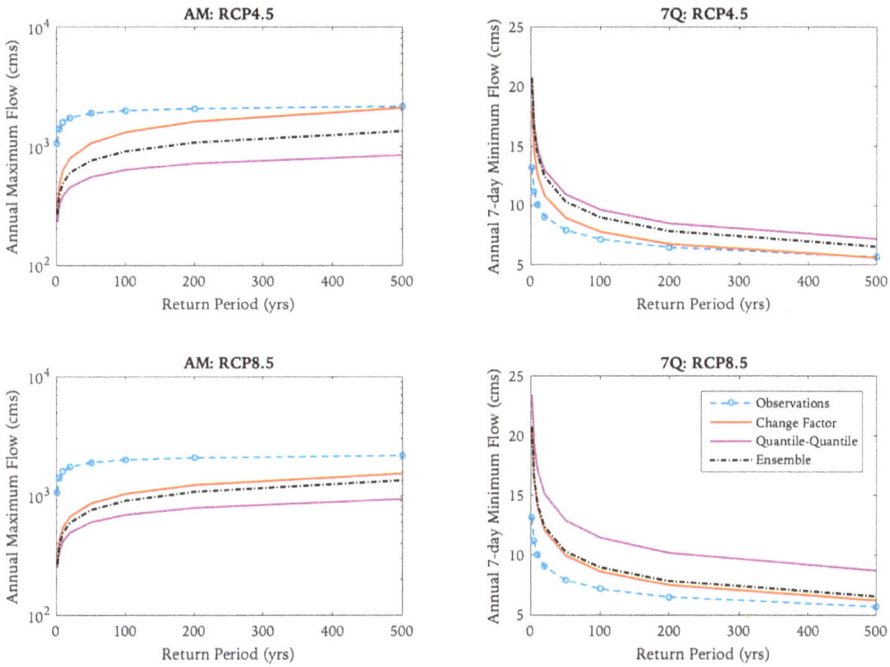

Figure 4. Estimations of future drought indices (7Q) and design floods (AM) based on RCP4.5 and RCP8.5 constructed from prolonged corrected climate time series, compared to observed and ensemble series.

3.3.2. Spring Flow Variability

Increasing temperatures have a direct effect on the hydrology of the watershed by changing to the snowmelt-driven runoff that occurs in late winter/early spring. Shorter duration of snowpack and an increase in stream flow in the winter are projected, triggered by a warming winter, particularly in cold regions. This is predicted by most models and scenarios with some variations in timing and magnitude of the largest flow with more substantial warming occurring in the winter months. Figure 5 shows the disparity between the timing of the largest flow in the historical period and in the late-21st century. Between 1981 and 2010, the median date of the largest spring flow was March 12th (red dashed line in Figure 5b). The domination of most of the models and downscaling methods suggests that the peak spring flow will occur earlier, which is consistent with an earlier spring warming. As a result, the intensity of the annual spring flow at the outlet of the SNW is expected to significantly decrease by up to 60% (Figure 5a). The ensemble results suggest a 50% decrease in the quantity and the occurrence of the peak spring flow to be before the month of March.

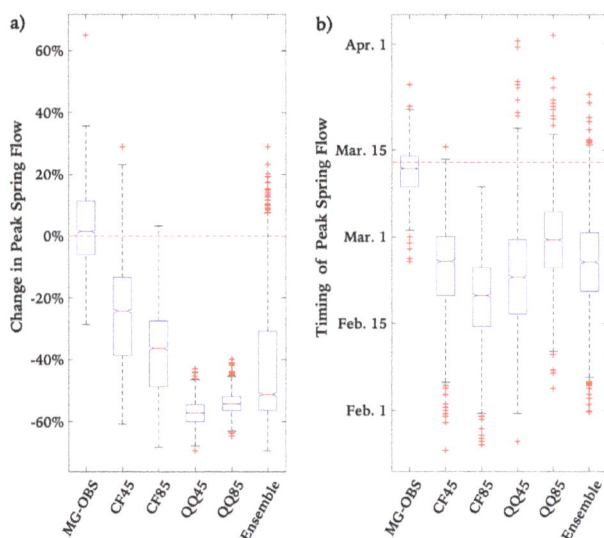

Figure 5. Estimations Projected changes in (**a**) the magnitude (%) and (**b**) timing of the peak spring flow event at the outlet of the SNW using two downscaling methods (Change Factor {CF}, and Quantile-Quantile method {QQ}) and forced with two climate scenarios (RCP4.5 {45} and RCP8.5 {85}). "Ensemble" represents all simulations combined and compared to the range of variability observed in the reference period (MG-OBS).

4. Discussion

The main contribution of this paper is to represent the variability of future outcomes through ensembles of realizations of climate time series that are converted into ensemble of impacts. The spread of these ensembles is a representation of the uncertainty in future climate and impacts. The uncertainty comes from the fact that there is no consensus approximation of climate response to a doubling of atmospheric carbon dioxide (or climate sensitivity) although the initial estimation by National Academy of Sciences in 1979 [65] to be in the range of 1.5–4.5 °C was still adopted by the recent reports of the Intergovernmental Panel on Climate Change (IPCC) [66,67]. Andrews et al. [68] suggested that the response of global temperature might be greater and more uncertain than previously estimated, specifically at the upper end. The ensemble also covers the range of variability observed during the baseline period. The spread of the ensemble depends on the climate model, emission scenario and the choice of downscaling methods. The following sections discuss the impacts of these factors on the results.

4.1. Comparison of Downscaling Methods

By comparing downscaling methods in Figures 3 and 5, it becomes apparent that the QQ technique outperformed the mean-based CF by correcting the distributions of RCM data rather than a single statistic. The cloud of points is denser (yellow contours) where the probability is higher. QQ outcomes have a tendency to consistent clouds representing different climate models and scenarios. Indeed, the convergence in the projected QQ results across RCMs with different parameterization and climate sensitivity may suggest a more refined and skillful representation of future climate and hydrological systems, especially in the estimation of extreme daily precipitation [42]. However, in the context of risk analysis, Daniels et al. [69] stated that one cannot utterly exclude an outlier scenario as it may characterize some underlying physical processes that were missed by other models or scenarios.

The mean-based downscaling method used in this study provides more diverse results (Figure 3). Results, however, are scattered, suggesting that the choice of the downscaling method will greatly affect the estimated impacts and thus adaptation strategies. In the absence of evidence that one method is better than another, the safest approach is to use all available methods. Several researchers or practitioners use one downscaling method without justification, neglecting that their results will be tainted by that choice. Research is needed to assess the relative credibility of downscaling methods in order to choose the most reliable approach.

4.2. Sensitivity Domains Assessment

Based on the probabilistic assessment of climate change, the 7-day mean annual minima and annual maxima indices suggested positive changes in the future period compared to the reference period. While low streamflow is a product of a complex combination of several factors in the physical processes, a possible reason of such projected changes in low flow is attributed to enhanced evapotranspiration (due to increased air temperature) from the wet-land surfaces of the watershed in summer [70], which leads to increased precipitation frequency through more intense convective storms and consequently increasing flows during the summer months. These projected changes of volume in low flows would be welcome especially in terms of their impacts on aquatic habitat and water quality.

Higher values of future low flows compared to the baseline period due to increased rainfall intensity in North America were reported by Shrestha et al. [71] based on SWAT simulations forced by data from ten GCMs from CMIP5 archives in response to RCP4.5 and RCP8.5 scenarios in a river located in southwest Ohio, USA. Similar conclusions of extremely low flow conditions to occur less frequently in the future were reported by modeling studies applied to different watersheds across Europe [72,73] and Asia [74,75]. In fact, these projections are also in agreement with previously reported findings of upward trends in observed low flow values across Canada [76,77]. For example, Khaliq et al. [78] investigated observed values of 1-, 7-, 15-, and 30-day annual low-flow indicators in some Canadian rivers and found significantly increasing trends in southern Ontario. Similarly, Moore et al. [79] found a statistically large positive trend in streamflow during low flow season in southwest British Columbia. Moreover, Novotny and Stefan [80] found 7-day low flows to be increasing at a significant rate in three major river basins of Minnesota during both the summer and winter. In general, both low- and high-flows have been investigated in various regional studies, with no consistent conclusions in terms of how they will be altered by climate change [21]. Rather, it is believed that such extremes are very localized and dependent on the selected GCMs and hydrological models; thus results cannot be simply generalized.

On the other hand, spring peak flow is generally a product of average air temperature and snowmelt runoff. Most of the ensemble data suggest an early occurrence of peak spring flow. This result is consistent with the finding of Gunawardhana and Kazama [72], who projected an early occurrence of peak spring flow in Italy by approximately 12 days during the 2080s in comparison to the baseline period. Also, it is supported by observed and projected changes in spring streamflow-timing across parts of western North America [81]. An advance in the timing of spring peak flow could negatively affect water supply, ecosystem, and reservoirs' storage management [81]. However, such change in timing implies a reduction in snowmelt runoff severity (and hence the magnitude of the largest daily mean flows in spring), and the economic impacts will be apparent particularly in Canada [82]. These findings can likely explain the projected reduction in annual maxima values in a mainly snowmelt dominated river, where typically the highest flow occurs in early spring each year caused by snowmelt-generated runoff. These results are also consistent with previous works done in some Canadian watersheds that detected the same behaviour in the past such as Zhang et al. [83].

When multiple RCMs and downscaling approaches are used, the best hydrological estimate can be constructed based on weighting of combined results of rainfall-runoff modelling. By adopting the versatile methodology considered herein, it is believed that the risk of extremes and their consequent losses under an uncertain climate can be significantly reduced, as risk cannot be completely eliminated.

Ideally, further improvement to the super-ensemble can be achieved by considering more stochastic weather generators, climate-change models, and downscaling methods to ensure a thorough evaluation. To overcome the high computational requirements to construct the super-ensemble, further research is needed to determine the optimal length of the chain containing realizations needed to represent observed data in climatic and hydrological modelling.

5. Conclusions

This paper proposed a novel approach to associate a credibility measure to the climate and sensitivity measure in the bottom-up approach by combining synthetically generated climate series with downscaled climate model output to populate the climate and sensitivity spaces. The large number of projections allowed us to quantify future uncertainties and provided better coverage of the risk space based on probabilistic assessment of climate change. Furthermore, the likelihood measure provides a more practical assessment of the plausibility of future risks to serve infrastructure design and allow more confidence in water-related management decisions. Ideally, the generation of such ensembles should contain—to a feasible extent—as many uncertainty elements as possible. Given that the choice of climate models, downscaling techniques, and weather generator affects the results, a super-ensemble of future climate series was used to derive flooding and drought indices, while bearing in mind the uncertainties inherent in the extreme value modelling under a changing climate at regional scales. The results of the modelled risk analysis on the South Nation Watershed located in Ontario, Canada, indicate a marginal increase in the 7-day low flow, whereas design floods will noticeably weaken. A shorter and warmer winter is expected to result in an earlier disappearance of accumulated winter snow cover, an early onset of snowmelt, and consequently an earlier and less intensive peak spring flow. While an application was centered in one pilot watershed, the methodology delivers new insights into hydrological processes under changing climate conditions in general and will be of interest for Canada and beyond. The need is also manifest for examining a broader range of hydrologic indicators. Finally, the broader natural uncertainty posed by climate change and land use, and their implicit disruptions to various agro-socioeconomic and water sectors at an operational level, may constitute an interesting and worthwhile extension of this study.

Author Contributions: A.A. and O.S. conceived and designed the study. A.A. collected the data, performed the computations, analyzed the results and wrote the original draft of the manuscript. O.S. supervised the research work and contributed to the interpretation of the results, and manuscript revision.

Funding: The research was partially funded by the National Sciences and Engineering Research Council discovert grant (RGPIN/ 2016-05094) to the second author.

Conflicts of Interest: The authors declare no conflict of interest.

References

1. Stocker, T.F.; Qin, D.; Plattner, G.-K.; Tignor, M.; Allen, S.K.; Boschung, J.; Nauels, A.; Xia, Y.; Bex, V.; Midgley, P.M. *Climate Change 2013: The Physical Science Basis. Intergovernmental Panel on Climate Change, Working Group I Contribution to the IPCC Fifth Assessment Report (AR5)*; Cambridge University Press: New York, NY, USA, 2013; pp. 159–254.
2. Manabe, S. Climate and the ocean circulation: I. The atmospheric circulation and the hydrology of the earth's surface. *Mon. Weather Rev.* **1969**, *97*, 739–774. [CrossRef]
3. Gleick, P.H. Climate change, hydrology, and water resources. *Rev. Geophys.* **1989**, *27*, 329–344. [CrossRef]
4. Whitfield, P.H.; Cannon, A.J. Recent variations in climate and hydrology in Canada. *Can. Water Resour. J.* **2000**, *25*, 19–65. [CrossRef]
5. Arora, V.K.; Boer, G.J. Effects of simulated climate change on the hydrology of major river basins. *J. Geophys. Res. Atmos.* **2001**, *106*, 3335–3348. [CrossRef]
6. Frich, P.; Alexander, L.V.; Della-Marta, P.M.; Gleason, B.; Haylock, M.; Tank, A.K.; Peterson, T. Observed coherent changes in climatic extremes during the second half of the twentieth century. *Clim. Res.* **2002**, *19*, 193–212. [CrossRef]

7. Bierkens, M.F.; Dolman, A.J.; Troch, P.A. (Eds.) *Climate and the Hydrological Cycle*; International Association of Hydrological Sciences: Wallingford, UK, 2008; Volume 8.

8. Panagoulia, D. Hydrological response of a medium-sized mountainous catchment to climate changes. *Hydrol. Sci. J.* **1991**, *36*, 525–547. [CrossRef]

9. Panagoulia, D. Impacts of GISS-modelled climate changes on catchment hydrology. *Hydrol. Sci. J.* **1992**, *37*, 141–163. [CrossRef]

10. Panagoulia, D.; Bárdossy, A.; Lourmas, G. Multivariate stochastic downscaling models generating precipitation and temperature scenarios of climate change based on atmospheric circulation. *Glob. Nest J.* **2008**, *10*, 263–272.

11. Pielke Sr, R.A.; Wilby, R.; Niyogi, D.; Hossain, F.; Dairuku, K.; Adegoke, J.; Suding, K. Dealing with complexity and extreme events using a bottom-up, resource-based vulnerability perspective. Extreme Events and Natural Hazards: The Complexity Perspective. *Geophys. Monogr. Ser.* **2012**, *196*, 345–359.

12. García, L.E.; Matthews, J.H.; Rodriguez, D.J.; Wijnen, M.; DiFrancesco, K.N.; Ray, P. *Beyond Downscaling: A Bottom-up Approach to Climate Adaptation for Water Resources Management*; World Bank Publications: Washington, DC, USA, 2014.

13. Brown, C.; Werick, W.; Leger, W.; Fay, D. A Decision-Analytic Approach to Managing Climate Risks: Application to the Upper Great Lakes. *J. Am. Water. Resour. Assoc.* **2011**, *47*, 524–534. [CrossRef]

14. Wilby, R.L. Adaptation: Wells of wisdom. *Nat. Clim. Chang.* **2011**, *1*, 302–303. [CrossRef]

15. Bhave, A.; Mishra, A.; Raghuwanshi, N. A combined bottom-up and top-down approach for assessment of climate change adaptation options. *J. Hydrol.* **2014**, *518*, 150–161. [CrossRef]

16. Culley, S.; Noble, S.; Yates, A.; Timbs, M.; Westra, S.; Mair, H.R.; Giuliani, M.; Castelletti, A. A bottom-up approach to identifying the operational adaptive capacity of water resources systems to a changing climate. *Water Resour. Res.* **2016**, *52*, 6751–6758. [CrossRef]

17. Alodah, A.; Seidou, O. The realism of Stochastic Weather Generators in Risk Discovery. *WIT Trans. Ecol. Environ.* **2017**, *220*, 239–249. [CrossRef]

18. Guo, D.; Westra, S.; Maier, H.R. Use of a scenario-neutral approach to identify the key hydro-meteorological attributes that impact runoff from a natural catchment. *J. Hydrol.* **2017**, *554*, 317–330. [CrossRef]

19. Seidou, O.; Alodah, A. From top-down to bottom-up approaches to risk discovery: a paradigm shift in climate change impacts and adaptation studies related to the water sector. In Proceedings of the Annual Conference of the Canadian Society for Civil Engineering (CSCE2018), Fredericton, NB, Canada, 13–16 June 2018. DM36-01-11.

20. Panagoulia, D. Artificial neural networks and high and low flows in various climate regimes. *Hydrol. Sci. J.* **2006**, *51*, 563–587. [CrossRef]

21. Sapač, K.; Medved, A.; Rusjan, S.; Bezak, N. Investigation of Low- and High-Flow Characteristics of Karst Catchments under Climate Change. *Water* **2019**, *11*, 925. [CrossRef]

22. Tharme, R.E. A global perspective on environmental flow assessment: Emerging trends in the development and application of environmental flow methodologies for rivers. *River Res. Appl.* **2003**, *19*, 397–441. [CrossRef]

23. Arthington, A.H.; Bunn, S.E.; Poff, N.L.; Naiman, R.J. The challenge of providing environmental flow rules to sustain river ecosystems. *Ecol. Appl.* **2006**, *16*, 1311–1318. [CrossRef]

24. Linnansaari, T.; Monk, W.A.; Baird, D.J.; Curry, R.A. Review of approaches and methods to assess Environmental Flows across Canada and internationally. *DFO Can. Sci. Advis. Secr. Res. Doc.* **2012**, *39*, 1–74.

25. Pastor, A.V.; Ludwig, F.; Biemans, H.; Hoff, H.; Kabat, P. Accounting for environmental flow requirements in global water assessments. *Hydrol. Earth Syst. Sci.* **2014**, *18*, 5041–5059. [CrossRef]

26. Chin, V.I.; Wang, K.T.; Vallery, O.J. *Water Resources of the South Nation River Basin*; Rep. No. 13; Ontario Ministry of the Environment, Water Resources Branch: Toronto, ON, Canada, 1980.

27. Alodah, A.; Seidou, O. The adequacy of stochastically generated climate time series for water resources systems risk and performance assessment. *Stoch. Environ. Res. Risk Assess.* **2019**, *33*, 253–269. [CrossRef]

28. Semenov, V.A. Structure of temperature variability in the high latitudes of the Northern Hemisphere. *Izv. Atmos. Ocean. Phys.* **2007**, *43*, 687–695. [CrossRef]

29. Lenderink, G.; Buishand, A.; Deursen, W.V. Estimates of future discharges of the river Rhine using two scenario methodologies: Direct versus delta approach. *Hydrol. Earth Syst. Sci.* **2007**, *11*, 1145–1159. [CrossRef]

30. Diaz-Nieto, J.; Wilby, R.L. A comparison of statistical downscaling and climate change factor methods: Impacts on low flows in the River Thames, United Kingdom. *Clim. Chang.* **2005**, *69*, 245–268. [CrossRef]

31. Hansen, C.H.; Goharian, E.; Burian, S. Downscaling precipitation for local-scale hydrologic modeling applications: Comparison of traditional and combined change factor methodologies. *J. Hydrol. Eng.* **2017**, *22*, 04017030. [CrossRef]

32. Harris, C.N.P.; Quinn, A.D.; Bridgeman, J. The use of probabilistic weather generator information for climate change adaptation in the UK water sector. *Meteorol. Appl.* **2014**, *21*, 129–140. [CrossRef]

33. Lafon, T.; Dadson, S.; Buys, G.; Prudhomme, C. Bias correction of daily precipitation simulated by a regional climate model: A comparison of methods. *Int. J. Climatol.* **2013**, *33*, 1367–1381. [CrossRef]

34. Luo, M.; Liu, T.; Frankl, A.; Duan, Y.; Meng, F.; Bao, A.; De Maeyer, P. Defining spatiotemporal characteristics of climate change trends from downscaled GCMs ensembles: How climate change reacts in Xinjiang, China. *Int. J. Climatol.* **2018**, *38*, 2538–2553. [CrossRef]

35. Mpelasoka, F.S.; Chiew, F.H.S. Influence of rainfall scenario construction methods on runoff projections. *J. Hydrometeorol.* **2009**, *10*, 1168–1183. [CrossRef]

36. Teutschbein, C.; Seibert, J. Bias correction of regional climate model simulations for hydrological climate-change impact studies: Review and evaluation of different methods. *J. Hydrol.* **2012**, *456*, 12–29. [CrossRef]

37. Van Roosmalen, L.; Sonnenborg, T.O.; Jensen, K.H.; Christensen, J.H. Comparison of hydrological simulations of climate change using perturbation of observations and distribution-based scaling, Soil Science Society of America. *Vadose Zone J.* **2011**, *10*, 136–150. [CrossRef]

38. Hayhoe, K.A. A Standardized Framework for Evaluating the Skill of Regional Climate Downscaling Techniques. Ph.D. Thesis, University of Illinois at Urbana-Champaign, Champaign County, IL, USA, 2010.

39. Seidou, O.; Ramsay, A.; Nistor, I. Climate change impacts on extreme floods II: Improving flood future peaks simulation using non-stationary frequency analysis. *Nat. Hazards* **2012**, *60*, 715–726. [CrossRef]

40. Maraun, D. Bias correcting climate change simulations-a critical review. *Curr. Clim. Chang. Rep.* **2016**, *2*, 211–220. [CrossRef]

41. Themβel, M.J.; Gobiet, A.; Leuprecht, A. Empirical statistical downscaling and error correction of daily precipitation from regional climate models. *Int. J. Climatol.* **2010**, *31*, 1530–1544. [CrossRef]

42. Sarr, M.A.; Seidou, O.; Tramblay, Y.; El Adlouni, S. Comparison of downscaling methods for mean and extreme precipitation in Senegal. *J. Hydrol. Reg. Stud.* **2015**, *4*, 369–385. [CrossRef]

43. Angelina, A.; Gado Djibo, A.; Seidou, O.; Seidou Sanda, I.; Sittichok, K. Changes to flow regime on the Niger River at Koulikoro under a changing climate. *Hydrol. Sci. J.* **2015**, *60*, 1709–1723. [CrossRef]

44. Wilby, R.L.; Fowler, H.J. *Regional Climate Downscaling: Modelling the Impact of Climate Change on Water Resources*; Fung, F., Lopez, A., New, M., Eds.; Wiley-Blackwell: Chichester, West Sussex, UK; Hoboken, NJ, USA, 2011; ISBN 978-1-4051-9671-0.

45. Wilks, D.S. Multi-site generalization of a daily stochastic precipitation model. *J. Hydrol.* **1998**, *210*, 178–191. [CrossRef]

46. Chen, J.; Brissette, F.P.; Zhang, X.J. A multi-site stochastic weather generator for daily precipitation and temperature. *Trans. ASABE* **2014**, *57*, 1375–1391.

47. Abbaspour, K.C.; Yang, J.; Maximov, I.; Siber, R.; Bogner, K.; Mieleitner, J.; Srinivasan, R. Modelling hydrology and water quality in the pre-alpine/alpine Thur watershed using SWAT. *J. Hydrol.* **2007**, *333*, 413–430. [CrossRef]

48. Neitsch, S.L.; Arnold, J.G.; Kiniry, J.R.; Williams, J.R. *Soil and Water Assessment Tool Theoretical Documentation Version 2009*; Texas Water Resources Institute: College Station, TX, USA, 2011.

49. Srinivasan, R.; Arnold, J.G. Integration of a basin-scale water quality model with GIS. *Water Resour. Bull.* **1994**, *30*, 453–462. [CrossRef]

50. Arnold, J.G.; Srinivasan, R.; Muttiah, R.S.; Williams, J.R. Large area hydrologic modeling and assessment part I: Model development 1. *J. Am. Water. Resour. Assoc.* **1998**, *34*, 73–89. [CrossRef]

51. White, K.L.; Chaubey, I. Sensitivity analysis, calibration, and validations for a multisite and multivariable SWAT model. *J. Am. Water. Resour. Assoc.* **2005**, *41*, 1077–1089. [CrossRef]

52. Tuppad, P.; Douglas-Mankin, K.R.; Lee, T.; Srinivasan, R.; Arnold, J.G. Soil and Water Assessment Tool (SWAT). hydrologic/water quality model: Extended capability and wider adoption. *Trans. ASABE* **2011**, *54*, 1677–1684. [CrossRef]

53. Arnold, J.G.; Moriasi, D.N.; Gassman, P.W.; Abbaspour, K.C.; White, M.J.; Srinivasan, R.; Santhi, C.; Harmel, R.D.; Van Griensven, A.; Van Liew, M.W.; et al. SWAT: Model use, calibration, and validation. *Trans. ASABE* **2012**, *55*, 1491–1508. [CrossRef]

54. Santhi, C.; Arnold, J.; Williams, J.; Dugas, W.; Srinivasan, R.; Hauck, L. Validation of the SWAT model on a large river basin with point and nonpoint sources 1. *J. Am. Water. Resour. Assoc.* **2001**, *37*, 1169–1188. [CrossRef]

55. Tan, M.L.; Gassman, P.W.; Srinivasan, R.; Arnold, J.G.; Yang, X. A Review of SWAT Studies in Southeast Asia: Applications, Challenges and Future Directions. *Water* **2019**, *11*, 914. [CrossRef]

56. Wilby, R.L.; Charles, S.P.; Zorita, E.; Timbal, B.; Whetton, P.; Mearns, L.O. *Guidelines for Use of Climate Scenarios Developed from Statistical Downscaling Methods*; Supporting material of the Intergovernmental Panel on Climate Change; DDC of IPCC TGCIA: Geneva, Switzerland, 2004.

57. Brown, C.; Ghile, Y.; Laverty, M.; Li, K. Decision scaling: Linking bottom up vulnerability analysis with climate projections in the water sector. *Water Resour. Res.* **2012**, *48*, 9537. [CrossRef]

58. Jenkinson, A.F. The frequency distribution of the annual maximum (or minimum) values of meteorological elements. *Q. J. R. Meteorol. Soc.* **1955**, *81*, 158–171. [CrossRef]

59. Das, S.; Millington, N.; Simonovic, S.P. Distribution choice for the assessment of design rainfall for the city of London (Ontario, Canada) under climate change. *Can. J. Civ. Eng.* **2013**, *40*, 121–129. [CrossRef]

60. Huntington, T.G.; Hodgkins, G.A.; Dudley, R.W. Historical trend in river ice thickness and coherence in hydroclimatological trends in Maine. *Clim. Chang.* **2003**, *61*, 217–236. [CrossRef]

61. Karl, T.R.; Groisman, P.Y.; Knight, R.W.; Heim, R.R., Jr. Recent variations of snow cover and snowfall in North America and their relation to precipitation and temperature variations. *J. Clim.* **1993**, *6*, 1327–1344. [CrossRef]

62. Botev, Z.I.; Grotowski, J.F.; Kroese, D.P. Kernel density estimation via diffusion. *Ann. Stat.* **2010**, *38*, 2916–2957. [CrossRef]

63. Liew, M.W.; Veith, T.L.; Bosch, D.D.; Arnold, J.G. Suitability of SWAT for the conservation effects assessment project: A comparison on USDA-ARS experimental watersheds. *J. Hydrol. Eng.* **2007**, *12*, 173–189. [CrossRef]

64. Moriasi, D.N.; Arnold, J.G.; Van Liew, M.W.; Bingner, R.L.; Harmel, R.D.; Veith, T.L. Model evaluation guidelines for systematic quantification of accuracy in watershed simulations. *Trans. ASABE* **2007**, *50*, 885–900. [CrossRef]

65. National Academy of Sciences. *Carbon Dioxide and Climate: A Scientific Assessment*; National Academy of Sciences: Washington, DC, USA, 1979.

66. IPCC. *2007: Climate Change 2007: Synthesis Report. Contribution of Working Groups I, II and III to the Fourth Assessment Report of the Intergovernmental Panel on Climate Change*; Cre Writing Team, Pachauri, R.K., Reisinger, A., Eds.; IPCC: Geneva, Switzerland, 2007; 104p.

67. IPCC. *2014: Climate Change 2014: Impacts, Adaptation, and Vulnerability. Part A: Global and Sectoral Aspects. Contribution of Working Group II to the Fifth Assessment Report of the Intergovernmental Panel on Climate Change*; Field, C.B., Barros, V.R., Dokken, D.J., Mach, K.J., Mastrandrea, M.D., Bilir, T.E., Chatterjee, M., Ebi, K.L., Estrada, Y.O., Genova, R.C., et al., Eds.; Cambridge University Press: Cambridge, UK; New York, NY, USA, 2014; 1132p.

68. Andrews, T.; Gregory, J.M.; Paynter, D.; Silvers, L.G.; Zhou, C.; Mauritsen, T.; Titchner, H. Accounting for changing temperature patterns increases historical estimates of climate sensitivity. *Geophys. Res. Lett.* **2018**, *45*, 8490–8499. [CrossRef]

69. Daniels, A.E.; Morrison, J.F.; Joyce, L.A.; Crookston, N.L.; Chen, S.C.; McNully, S.G. *Climate Projections FAQ*; General Technical Report; Department of Agriculture, Forest Service, Rocky Mountain Research Station: Fort Collins, CO, USA, 2012; pp. 1–32.

70. Abdel-Fattah, S.; Krantzberg, G. A review: Building the resilience of Great Lakes beneficial uses to climate change. *Sustain. Water Qual. Ecol.* **2014**, *3*, 3–13. [CrossRef]

71. Shrestha, S.; Sharma, S.; Gupta, R.; Bhattarai, R. Impact of global climate change on stream low flows: A case study of the great Miami river watershed, Ohio, USA. *Int. J. Agric. Biol. Eng.* **2019**, *12*, 84–95. [CrossRef]

72. Gunawardhana, L.N.; Kazama, S. A water availability and low-flow analysis of the Tagliamento River discharge in Italy under changing climate conditions. *Hydrol. Earth Syst. Sci.* **2012**, *16*, 1033–1045. [CrossRef]

73.	Laaha, G.; Parajka, J.; Viglione, A.; Koffler, D.; Haslinger, K.; Schöner, W.; Zehetgruber, J.; Blöschl, G. A three-pillar approach to assessing climate impacts on low flows. *Hydrol. Earth Syst. Sci.* **2016**, *20*, 3967–3985. [CrossRef]

74.	Gain, A.K.; Immerzeel, W.W.; Sperna Weiland, F.C.; Bierkens, M.F.P. Impact of climate change on the stream flow of the lower Brahmaputra: Trends in high and low flows based on discharge-weighted ensemble modelling. *Hydrol. Earth Syst. Sci.* **2011**, *15*, 1537–1545. [CrossRef]

75.	Tian, Y.; Xu, Y.P.; Ma, C.; Wang, G. Modeling the impact of climate change on low flows in Xiangjiang River Basin with Bayesian averaging method. *J. Hydrol. Eng.* **2017**, *22*, 04017035. [CrossRef]

76.	Ehsanzadeh, E.; Adamowski, K. Detection of trends in low flows across Canada. *Can. Water Resour. J.* **2007**, *32*, 251–264. [CrossRef]

77.	Yue, S.; Pilon, P.; Phinney, B.O.B. Canadian streamflow trend detection: Impacts of serial and cross-correlation. *Hydrol. Sci. J.* **2003**, *48*, 51–63. [CrossRef]

78.	Khaliq, M.N.; Ouarda, T.B.; Gachon, P.; Sushama, L. Temporal evolution of low-flow regimes in Canadian rivers. *Water Resour. Res.* **2008**, *44*. [CrossRef]

79.	Moore, R.D.; Allen, D.M.; Stahl, K. *Climate Change and Low Flows: Influences of Groundwater and Glaciers*; Final Report for Climate Change Action Fund Projec A; Hydrology Applications Group, Environment Canada: Vancouver, BC, Canada, 2007; Volume 875, p. 211.

80.	Novotny, E.V.; Stefan, H.G. Stream flow in Minnesota: Indicator of climate change. *J. Hydrol.* **2007**, *334*, 319–333. [CrossRef]

81.	Stewart, I.T.; Cayan, D.R.; Dettinger, M.D. Changes in snowmelt runoff timing in western North America under abusiness as usual'climate change scenario. *Clim. Chang.* **2004**, *62*, 217–232. [CrossRef]

82.	Beltaos, S. Advances in river ice hydrology. *Hydrol. Process.* **2000**, *14*, 1613–1625. [CrossRef]

83.	Zhang, X.; Harvey, K.D.; Hogg, W.D.; Yuzyk, T.R. Trends in Canadian streamflow. *Water Resour. Res.* **2001**, *37*, 987–998. [CrossRef]

Article

A Comparative Analysis of the Historical Accuracy of the Point Precipitation Frequency Estimates of Four Data Sets and Their Projections for the Northeastern United States

Shu Wu [1,*], Momcilo Markus [2], David Lorenz [1], James R. Angel [2] and Kevin Grady [2]

[1] Nelson Institute Center for Climatic Research, University of Wisconsin–Madison, Madison, WI 53706, USA; david.lorenz@wisc.edu

[2] Prairie Research Institute, University of Illinois at Urbana-Champaign, Champaign, IL 61801, USA; mmarkus@illinois.edu (M.M.); jimangel@illinois.edu (J.R.A.); kagrady2@illinois.edu (K.G.)

* Correspondence: swu33@wisc.edu; Tel.: +1-608-209-8688

Received: 2 May 2019; Accepted: 14 June 2019; Published: 19 June 2019

Abstract: Many studies have projected that as the climate changes, the magnitudes of extreme precipitation events in the Northeastern United States are likely to continue increasing, regardless of the emission scenario. To examine this issue, we analyzed observed and modeled daily precipitation frequency (PF) estimates in the Northeastern US on the rain gauge station scale based on both annual maximum series (AMS) and partial duration series (PDS) methods. We employed four Coupled Model Intercomparison Project Phase 5 (CMIP5) downscaled data sets, including a probabilistic statistically downscaled data set developed specifically for this study. The ability of these four data sets to reproduce the observed features of historical point PF estimates was compared, and the two with the best historical accuracy, including the newly developed probabilistic data set, were selected to produce projected PF estimates under two CMIP5-based emission scenarios, namely Representative Concentration Pathway 4.5 (RCP4.5) and Representative Concentration Pathway 8.5 (RCP8.5). These projections indeed demonstrate a likely increase in PF estimates in the Northeastern US with noted differences in magnitudes and spatial distributions between the two data sets and between the two scenarios. We also quantified how the exceedance probabilities of the historical PF estimate values are likely to increase under each scenario using the two best performing data sets. Notably, an event with a current exceedance probability of 0.01 (a 100-year event) may have an exceedance probability for the second half of the 21st century of ≈0.04 (a 27-year event) under the RCP4.5 scenario and ≈0.05 (a 19-year event) under RCP8.5. Knowledge about the projected changes to the magnitude and frequency of heavy precipitation in this region will be relevant for the socio-economic and environmental evaluation of future infrastructure projects and will allow for better management and planning decisions.

Keywords: frequency estimates; downscaling; future projections; RCP4.5; RCP8.5; changing of exceedance; Northeastern US

---

1. Introduction

The time series of extreme precipitation events in the United States are not stationary [1], as the statistics describing them have been observed to change over the past several decades [2]. They will likely continue to change in the future as the climate changes (e.g., [3]). In the Northeastern US, in particular, extreme precipitation events have increased the most dramatically compared with any other region in the United States in the past several decades [4–7]. These changes then need to be investigated so that design standards can be met when constructing infrastructure that requires information about precipitation extremes. Many metrics could be used to describe the features of these extreme precipitation events, among which precipitation frequency (PF) analysis was selected for this study, as it is particularly important to hydrology design [8]. The goal of PF analysis is to obtain a useful estimate of a quantile for a relevant return period [9,10].

There are several methods of estimating future PFs. One method is to fit a nonstationary PF model to the observational data, such as a generalized extreme value (GEV) distribution, whose location, scale, and shape parameters are functions of time [11,12]. Consequently, the PF estimates are also functions of time with this method; substituting future dates into these PF functions results in future PF estimates. This method uses only historical observation data to obtain future PF estimates, so it may also be referred to as an observation extrapolation method. However, a key uncertainty of this method is that the linear trend assumption [12] may not hold in the future.

Another method is to use the output of climate models under different emission scenarios [2,13,14]. Climate models have demonstrated the ability to reproduce many observational features [15]; however, these models still present many uncertainties [16]. One of these uncertainties is caused by their focus mainly on larger climatic features on the continental or global scale. However, many hydrologic applications require smaller scales for which comparing model output with observations often involves a so-called downscaling process since the model resolutions are too coarse to be compared directly with the finer station observations. This downscaling process from the large to local scale, either done statistically [17,18] or dynamically [19,20], adds another layer of uncertainty when interpreting the results.

In this paper, we use model output to analyze point PF estimates for a large number of individual rain gauge stations throughout the Northeastern US. We compute PF estimates at each station for the historical period of 1960–2005 using each of the four downscaled Coupled Model Intercomparison Project Phase 5 (CMIP5) precipitation data sets, including three commonly used ones, and a fourth probabilistic data set developed for this study (Table 1). These modeled PF estimates are then compared with the observational PF estimates and each other. Models that are able to better reproduce historical data are assumed to be more accurate when making future projections [2,21,22]. The goal is then to find the data sets with the best overall historical accuracies to produce projected PF estimates and to quantify how the exceedance probabilities of the historical PF estimates are likely to change in the future under certain emission scenarios. These results will examine and highlight the need to reevaluate the existing PF estimates in the light of climate change.

This paper is organized as follows. We introduce the data sets and methodologies we will employ in Section 2, followed by comparisons between the observed and downscaled model data for the historical period in Section 3. Informed by these results, we select the best data sets and use them to present possible changes to the PF estimates in the future under two different Representative Concentration Pathways (RCPs), namely RCP4.5 and RCP8.5, in Section 4. The results in Section 4 demonstrate that PF estimates are very likely to increase in the future, suggesting that the exceedance probabilities of the historical PF estimates are also likely to increase. These changes are discussed in Section 5. Section 6 summarizes the results and provides conclusions and further discussions.

Table 1. Intergovernmental Panel on Climate Change (IPCC) models used to generate the downscale data sets.

Data Set	Models Used
UWPD (24)	ACCESS1-0; ACCESS1-3; CanESM2; CMCC-CESM; CMCC-CM CMCC-CMS; CNRM-CM5; CSIRO-Mk3-6-0; GFDL-CM3; GFDL-ESM2G GFDL-ESM2M; HadGEM2-CC; inmcm4; IPSL-CM5A-LR; IPSL-CM5A-MR IPSL-CM5B-LR; MIROC5; MIROC-ESM; MIROC-ESM-CHEM; MPI-ESM-LR MPI-ESM-MR; MRI-CGCM3; MRI-ESM1; NorESM1-M

Table 1. *Cont.*

Data Set	Models Used
LOCA (32)	ACCESS1-0; ACCESS1-3; bcc-csm1-1; bcc-csm1-1-m; CanESM2 CCSM4; CESM1-BGC; CESM1-CAM5; CMCC-CM; CMCC-CMS CNRM-CM5; CSIRO-Mk3-6-0; EC-EARTH; FGOALS-g2; GFDL-CM3 GFDL-ESM2G; GFDL-ESM2M; GISS-E2-H; GISS-E2-R; HadGEM2-AO HadGEM2-CC; HadGEM2-ES; inmcm4; IPSL-CM5A-LR; IPSL-CM5A-MR MIROC5; MIROC-ESM; MIROC-ESM-CHEM; MPI-ESM-LR; MPI-ESM-MR MRI-CGCM3; NorESM1-M
BCCAv2 (20)	ACCESS1-0; bcc-csm1-1; CanESM2; CCSM4; CESM1-BGC CNRM-CM5; CSIRO-Mk3-6-0; GFDL-CM3; GFDL-ESM2G; GFDL-ESM2M inmcm4; IPSL-CM5A-LR; IPSL-CM5A-MR; MIROC5; MIROC-ESM; MIROC-ESM-CHEM; MPI-ESM-LR; MPI-ESM-MR; MRI-CGCM3; NorESM1-M
NA-CORDEX (6)	CanESM2: {CRCM5(0.44°); RCA4(0.44°); CanRCM4(0.22°, 0.44°)}; EC-EARTH: {HIRHAMS(0.44°); RCA4(0.44°)}

Note: For the full model names, we refer to the CMIP5 website: https://cmip.llnl.gov/cmip5/docs/CMIP5_modeling_groups.docx.

2. Data and Methodologies

2.1. The Observational and Downscaled Model Data Sets

We examined observational data collected by the National Oceanic and Atmospheric Administration (NOAA) and used in the development of Atlas 14 [23]. The original data set included observations from 1218 rain gauge stations in the Northeastern US, of which 753 stations had a temporal coverage of over 80% for the period of 1960–2005 and were thus used in this study (Figure 1). Due to the limitations on the scope of this paper, we refer the reader to the published report [23] for more detailed quality control information about these data. The 1960–2005 period was selected for our analysis in order to have consistent comparisons with climate model outputs.

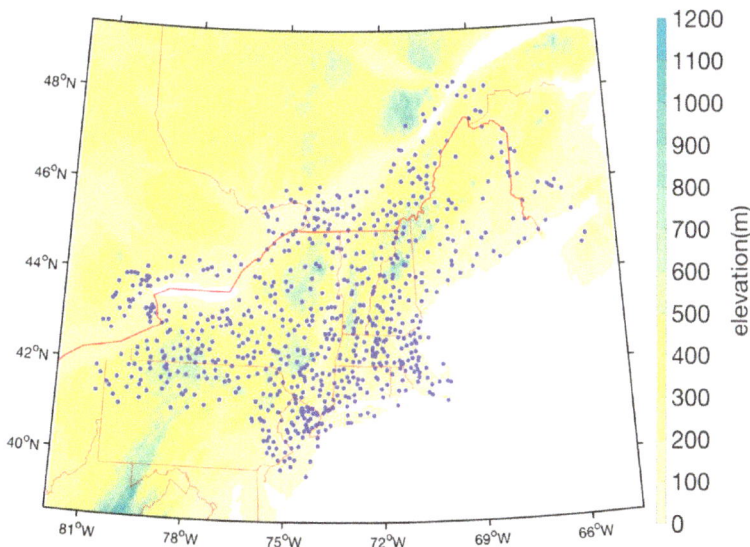

Figure 1. A map of the 753 selected stations in the Northeastern US. Shading represents elevation.

A large fraction of the study area lies along the northern Appalachian Mountains (Figure 1), which separate the coastal plain from the central lowlands. Local maxima of annual precipitation occur along the coast and higher elevations while local minima occur along the northern and western interior lowlands [4].

In addition to this observational data, four downscaled CMIP5 model data sets (Table 1) were analyzed. The first data set is a newly developed statistically downscaled product based on CMIP5 model output, hereafter referred to as the UWPD (University of Wisconsin Probabilistic Downscaling) data set [17]. This downscaled data set covers the Central and Eastern United States and Southern Canada at a 0.1 × 0.1 degree resolution. The downscaling methodology is fundamentally probabilistic [17]: The large-scale variables in a climate model do not determine the precise values of the downscaled variables. Instead, the large scale determines the likelihood of potential values of the downscaled variables; i.e., the large-scale variables determine the probability density function (PDF) of the downscaled variables. In the UWPD data, parametric PDFs are used, where the parameters of the distribution depend on the large-scale predictors. The simplest example of this approach is the use of ordinary least squares regression (OLS) to predict y from x. If the assumptions of OLS hold, then the expected value of y is Gaussian with a mean of the form ax + b (a and b are constants) and constant variance. In the UWPD data, the precipitation downscaling is divided into two components: (1) Estimate the probability that a day has precipitation or not (i.e., a binary variable), and (2) estimate the precipitation amount if the day is "wet." Logistic regression is used for the precipitation occurrence (first component), while a generalized gamma distribution [24] with parameters conditioned on the large scale is used for the precipitation amount (second component). Logistic regression was used because it is the appropriate regression analysis for binary data. Several potential link functions were explored for logistic regression, including the probit and several skewed link functions. It was found that the standard logit link function fit the observed data best. For the precipitation amount, the exponential, gamma, Weibull, and generalized exponential distributions were insufficiently flexible to capture the precipitation extremes. Therefore, the generalized gamma distribution, which has two shape parameters instead of one or zero, was used. The generalized gamma distribution includes the gamma and the Weibull distribution as special cases. Note, a single generalized gamma distribution is not used for all days, instead the parameters of the generalized gamma distribution depend on the daily varying large-scale atmospheric conditions. The precipitation amount methodology is very similar to that of the gamma distribution approach that Kirchmeier et al. [25] used for wind speed. The station data used to train the statistical models consist of daily precipitation accumulation from the National Weather Service (NWS)'s Cooperative Observer Program (1950–2009) and Environment Canada's Canadian Daily Climate Data (1950–2007) for approximately 4000 stations in the United States and Canada. Objective methods are applied to correct the hour of observation and remove stations with large biases in wet-day frequency [26], which results in the removal of 21% of stations. Only stations with at least 30 years of data and fewer than five missing days per month are included.

The large-scale atmospheric state for training the statistical models was obtained from the National Centers for Environmental Prediction (NCEP)–National Center for Atmospheric Research (NCAR) reanalysis [27]. The large-scale predictors include modeled precipitation itself, as well as variables involving moisture, wind, and static stability [17]. The large-scale fields were incorporated as predictors using the methodology in Kirchmeier et al. [25]. All statistical fits were cross-validated and only predictors with skill on independent data were used. To account for seasonality [4], separate fits were performed for each calendar month.

The time varying PDF parameters at each station that result from the above methodology were then interpolated in space [17]. The PDF parameters vary relatively slowly in space because they are determined from the large-scale atmospheric state; therefore, the interpolation used here results in almost no change in the statistics of the extreme events. Interpolating precipitation values, on the other hand, results in large reductions in precipitation variances due to the small-scale variability of precipitation. To generate "standard precipitation data" from the PDFs, random numbers are drawn from the PDFs to generate a "realization" of the small scale given the large scale. Such realizations are obviously not unique and are instead infinite in number and all equally likely. The random nature of the data accounts for the fact that for each particular large-scale state, there are multiple possible evolutions of the small-scale atmosphere that are consistent with this large-scale state. For example, the large-scale atmosphere on a particular day may be favorable for the development of small-scale thundershowers, but any individual point may have only a 50% probability of actually being hit by one of these storms. When generating realizations, the random numbers are correlated in space and time so that the spatial and temporal correlations of the downscaled variables are similar to observations. This is achieved by forcing an advection-diffusion model with independent random numbers and then using the output of the advection-diffusion model for drawing realizations from the PDFs. The winds at 3000 m altitude are used for the horizontal winds in the advection-diffusion model. The CMIP3 version of these data has been shown to have good skill [14,17,18,28]. For further information about this data set, we refer the reader to the website [29].

The second data set is the Localized Constructed Analogs (LOCAs) statistically downscaled data set [30]. These LOCA data have been widely used in government reports, such as the US Global Change Research Program (USGCRP, [31]). We used the downscaled results from 32 models from this data set (Table 1), each with a 1/16-degree resolution (≈6 km by 6 km). One simulation of each model was selected to form an ensemble of 32 members. For further information about this data set, we refer the reader to the LOCA website [32].

The third data set is the Bias-Correction and Constructed Analogs data (BCCAv2, [33]). The BCCAv2 data set has an ensemble size of 20 (one simulation of each model, Table 1) with 1/8-degree resolutions. For further information about this data set, we refer the reader to the website [34].

The final data set is the North America Coordinated Regional Downscaling Experiment data (NA-CORDEX, [35]). Although the previous three data sets were statistically downscaled (local variables are approximated from the large-scale predictor fields using statistical relationships), NA-CORDEX is instead dynamically downscaled. The data are generated by using the output of global models as lateral boundary and initial conditions to run high-resolution regional models. We selected downscaled data from two global models, namely CanESM2 and EC-EARTH, since the outputs of these two models are available across the historical, RCP4.5, and RCP8.5 periods. The CanESM2 model is paired with three regional models, namely CRCM5, RCA4, and CanRCM4, and EC-EARTH is paired with two regional models, namely RCA4 and HIRHAMS. All the data have a resolution of 0.44 degrees with the CanESM2 and CanRCM4 pairing having an extra simulation at a resolution of 0.22 degrees. This results in an ensemble of six members (Table 1). For further information about this data set, we refer the reader to the website [36].

The NA-CORDEX data set has not been completely generated yet, and our results (see Section 3.2) show that the BCCAv2 data set has relatively larger biases compared to the other data sets. Thus, our analysis will be mainly focused on the UWPD and LOCA data. However, the BCCAv2 and NA-CORDEX data also show some potential to perform well, especially the NA-CORDEX data; therefore, we provided some comparison results for these data as well for completeness. It should also be noted that the BCCAv2 and NA-CORDEX data are relatively coarse in resolution. This may lead to larger representation errors when comparing with point precipitation observations. Therefore, we applied an inverse of the empirical area reduction factors [14,37], producing 1.05 for the BCCAv2 data, 1.09 for the 0.22 degrees NA-CORDEX data, and 1.16 for the 0.44 degrees NA-CORDEX data.

2.2. Methodologies

Similar to NOAA Atlas 14 [23,38], our paper presents PF estimate results based on the annual maximum series (AMS) and partial duration series (PDS) of daily precipitation data. AMS and PDS are both widely used to carry out PF analysis in hydrology [39]. AMS are generated based on the maximum daily rainfall of each year, so the typical length of an AMS equals the number of years in the record. All of the values except for the one largest in a year are eliminated in calculating the PF estimates, which could potentially exclude some large events if two or more occur in a given year. In contrast, PDS are generated by retaining the values over a preselected threshold [40]. Therefore, the length of a PDS series is determined by the threshold, which may vary dramatically across different stations. Generally, the threshold is selected such that, on average, several events (2–4) occur per year, meaning the lengths of PDS are usually several times larger than those of the AMS for the same station and employ more observational data.

Because PDS use more data, one could presume that they give more significant PF estimates than AMS. However, more non-extreme values may be retained with PDS, which could lead to an underestimation of the final PFs, since PF analysis intrinsically focuses on extreme events. In practice, many products, such as NOAA Atlas 14, include results from both methods because AMS and PDS complement each other well and including both can also give more information on the uncertainties of the PF estimates. We too compared PF estimates from both AMS and PDS in our analysis to get a more complete understanding of how the downscaled data are able to simulate the observed features.

Calculating PF estimates using AMS versus PDS can be quite different in practice. For example, AMS are often associated with the generalized extreme value (GEV) distribution, while PDS are instead often associated with the generalized Pareto distribution (GPD) [40]. To make a direct comparison of AMS- and PDS-based PF results, we need to use consistent return periods for the two methods. However, the AMS return periods often need to be modified in order to do so since the AMS method tends to underestimate PF estimates for shorter return periods when compared with the PDS method [8]. For a given PDS return period, the adjusted AMS return period can be calculated as [41]:

$$T(AMS) = \left(1 - e^{-\frac{1}{T(PDS)}}\right)^{-1}. \tag{1}$$

For example, a return period of 2 years for a PDS may be equated to a return period of 2.54 years for an AMS. In this paper, we used "T = [2 5 10 25 50 100 200 500 1000]" years for the PDS return periods and "T = [2.54 5.52 10.51 25.50 50.50 100 200 500 1000]" years for the corresponding AMS return periods so that our PDS- and AMS-based PF estimates could be compared directly.

A more proper way to describe PF estimates associated with AMS might be to use an annual exceedance probability (AEP) instead of return periods [23]. Assuming the relationship holds that the AEP is the reciprocal of the return period for an AMS, we still used return periods for the convenience of comparison in this paper. For more about exceedance probabilities, see Section 5.

To generate a PDS, a threshold needs to be determined first. Although some methods are used to select the PDS threshold objectively, such as mean residual life plots [40], checking these plots for all stations to select the proper thresholds is virtually impossible. As a compromise, we examined three different methods of selecting thresholds in order to cover the associated uncertainties. For the first method, we used an iteration scheme to select a PDS threshold such that each year, on average, has two exceedance events. We treated consecutive exceedance days as one event, retaining only the largest value of the event. The second method is similar to the first, but the PDS threshold was lowered such that there are three events per year on average. For the third method, we used the 98% quantile level of non-zero precipitation values as the PDS threshold. In this paper, these three methods are labeled as PDS2, PDS3, and PDS qt98. The resulting thresholds for each station are shown in Figure 2a. This figure shows that, for most stations, PDS3 uses the lowest threshold and PDS2 uses the highest, with PDS qt98 falling in the middle. Different thresholds lead to different lengths of the base time series for PDS PF analysis (Figure 2b). As expected, the lengths of PDS2 and PDS3 are just double and triple, respectively, the number of analysis years, or 46 in our case (1960–2005), while the length of the PDS qt98 series varies by station. Please note that the stations are ordered by their location along the *x*-axis in this figure. The station order appears to be relatively arbitrary but has some tendency to cluster adjacent stations together. However, these patterns were not examined in this study.

Once we obtained the AMS and PDS for each station, GEV and GPD distributions, respectively, were fitted to the data to get the relevant estimated parameters by using the maximum likelihood method. PF estimates were then calculated using exceedance probabilities ([40], also see Section 5 of this paper).

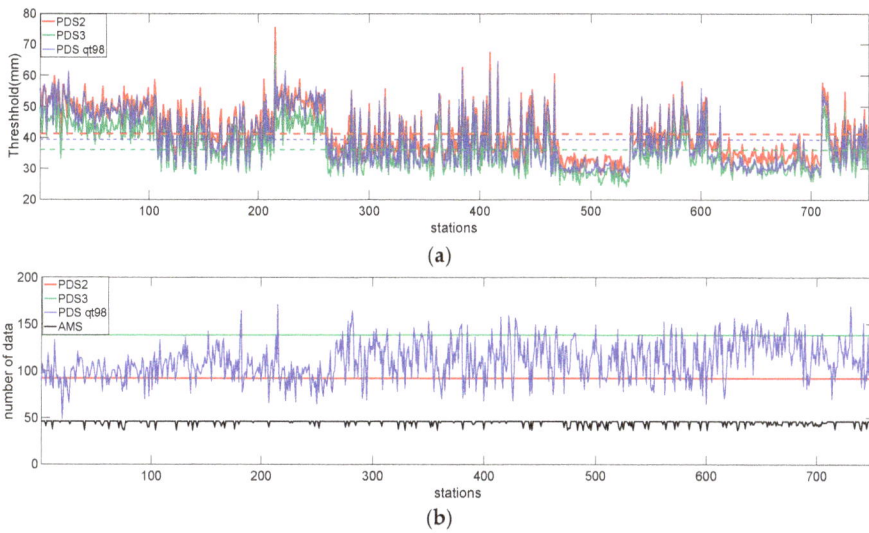

Figure 2. PDS thresholds (**a**) and length of data (**b**) used for PF analysis for each station. Red represents the results for PDS2, green for PDS3, blue for PDS qt98, and black for AMS.

3. Evaluations of the Downscaled Model Data for the Historical Period of 1960–2005

In this section, we evaluate the performance of the downscaled model data in terms of reproducing the observed features of the mean AMS, PDS, and PF estimates for different return periods. We will focus on the UWPD and LOCA data; some results for the BCCAv2 and NA-CORDEX data will also be provided at the end of this section. For this analysis, the closest model grid point to a rain gauge station was selected as its representation in the model.

3.1. Comparison of the Climatology of AMS and PDS of the Observed and Modeled Data

The 1960 to 2005 climatology (averages) of the observed AMS and PDS for each station are shown in the first row of Figure 3. The range is 40 to 120 mm, with larger values typically along the coast and smaller values inland. For the AMS (the first column of Figure 3), the differences between the UWPD data and these observed values are quite small in terms of difference ratios defined as (model − observation)/observation (Figure 3e). For over 80% of the stations, the difference ratio magnitudes are less than 10%, and there is no sign of systematic biases, as positive ratios are mixed together with negative ratios quite well. On the other hand, the LOCA data show systematic dry biases (negative ratios in Figure 3i) with difference ratio magnitudes of over 20% for most of the stations. For PDS, the climatology (Figure 3b–d) are in general smaller than for the AMS (Figure 3a), while the spatial distributions of the PDS are similar to each other. For every PDS threshold selection, the UWPD data generally have less than a 10% error compared with the observational results and no significant biases (Figure 3f–h), while the LOCA data generally have more than a 20% error and a systematic bias toward drier values, especially PDS qt98 (Figure 3j–l).

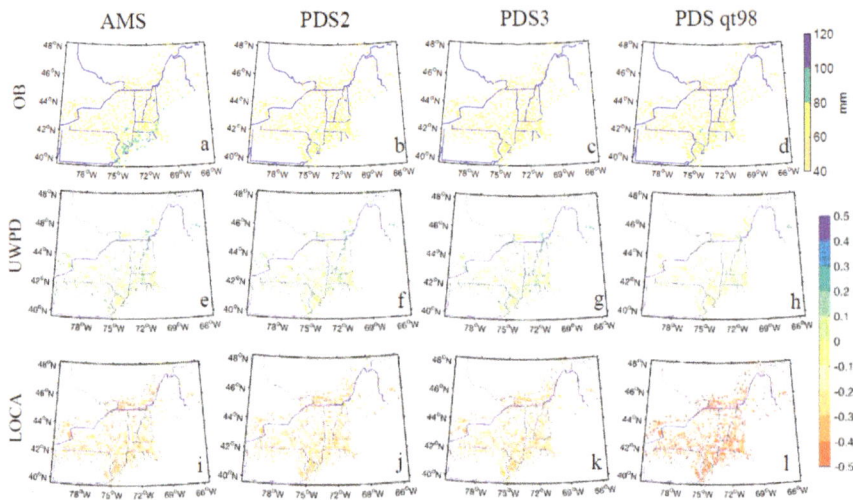

Figure 3. The 1960–2005 climatology of the AMS (**a**), PDS2 (**b**), PDS3 (**c**), and PDS qt98 (**d**), based on the observational data and the relative differences defined as (model – observation)/observation) for the UWPD (**e–h**) and LOCA (**i–l**) data.

3.2. Comparison of PF Estimates of the Observed and Modeled Data

In this section, we directly compare the observational PF estimates with those of the downscaled data to assess how well the model data may perform in estimating historical PF values. Figure 4 shows the PF estimate comparison results for the UWPD data, and Figure 5 shows the results for the LOCA data. The rows in each figure represent PF estimates for return periods of 2, 25, and 100 years, respectively. For each return period, we compared PF estimates from the downscaled data based on the AMS and each of the three PDS methods with their respective observed counterparts to get a more complete understanding. The y-axis represents the PF estimates calculated based on observations, and the x-axis represents the PF estimates based on the downscaled data. For each station, an ensemble mean (red dots) and standard deviation (blue bars) based on the PF estimates of the ensemble members of that downscaled data set can be calculated and used to estimate the mean value and associated uncertainty for that station. To assist in the comparison, lines of "y = x" are also plotted in red in each figure. More points above the line "y = x" mean the observed values are larger than the downscaled values, thus indicating dry biases in the models and vice versa.

Overall, the AMS and PDS demonstrate quite consistent features when compared. Results for the UWPD data (Figure 4) show that most of the points sit well around the line, "y = x", suggesting the UWPD PF estimates are close to the observed PF estimates, especially for shorter return periods. For a 2-year return period (Figure 4a–d), the UWPD PF estimates have a spatial correlation of 0.93 with the observations for all the AMS and PDS methods, suggesting that the spatial variations of the observational PF estimates are well represented by the UWPD data. The root mean square error (RMSE), which is around 4.5 mm on average, is also small compared to the observed values (40–90 mm). As the return periods become longer, the correlation coefficients decrease to ≈0.85 for the 25-year return period (Figure 4e–h) and ≈0.75 for the 100-year return period (Figure 4i–l). The RMSE increases to ≈13 mm for the 25-year return period and to ≈25 mm for the 100-year return period. The uncertainties (the length of the blue lines) also increase as the return periods become longer. It should be noted that for the extremely high and low PF estimates, the UWPD data indeed show some large errors manifested as dry biases for extremely high PF estimates and wet biases for extremely low PF estimates. Yet the number of these biased stations is small compared to the total number of stations, allowing us to conclude that the UWPD data overall capture the major features of the observed PF estimates. In contrast, the LOCA data consistently underestimate the PF estimate values across all the return periods (Figure 5), thus showing the same dry bias seen in Figure 3. The spatial correlation coefficients are quite high, however, indicating that the LOCA data can also capture the observed spatial variations relatively well. Standard deviations of the LOCA data (the length of the blue lines) are also smaller than for the UWPD data.

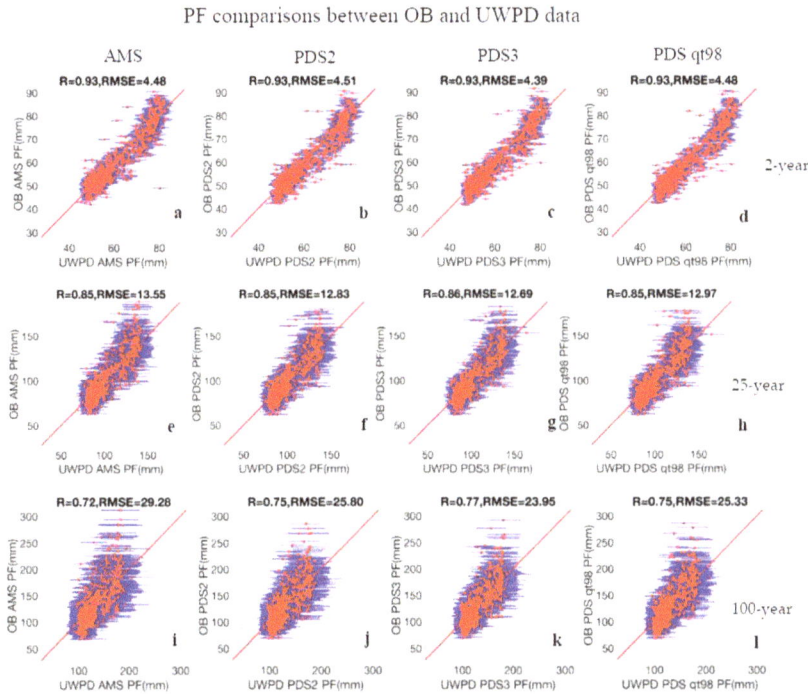

Figure 4. PF estimate comparisons between the observations and the UWPD data for return periods of 2 years (upper row), 25 years (middle row), and 100 years (lower row). The *y*-axis represents the PF estimates calculated based on observations, and the *x*-axis represents the PF estimates based on the downscaled data. For each station, an ensemble mean is represented by a red dot and the standard deviation is represented by a blue bar. Lines of "y = x" are also plotted for reference.

PF comparisons between OB and LOCA data

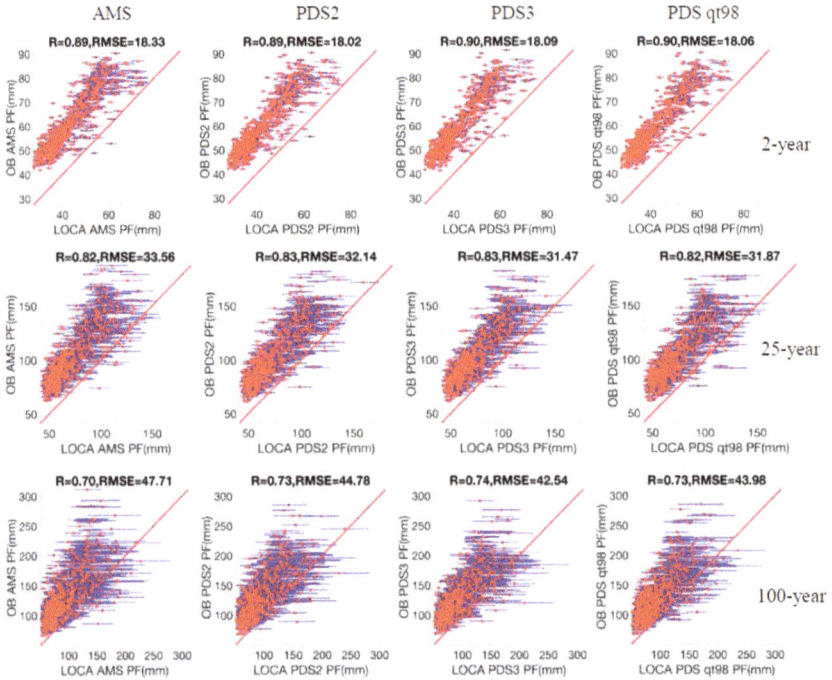

Figure 5. Similar to Figure 4 but for the LOCA data.

To summarize the comparison results, Figures 6 and 7 provide quantifications of skill, in terms of correlation coefficients and RMSE [8], respectively, for all four downscaled data sets across all return periods. The skills of the downscaled data sets decrease as return periods become longer. Quite consistently, we see that the UWPD data outperform the other data sets for all return periods. The LOCA data are the second best in terms of correlation skill, implying that although significant biases are present in the LOCA data, a simple bias correction may be adequate in making the LOCA data still useful for PF estimates. Interestingly, the only dynamically downscaled data set, NA-CORDEX, regardless of its relatively coarse resolution, also shows quite comparable skill for both the correlation coefficients and RMSE, although the uncertainties reflected by the ensemble spread are much larger than for all the statistically downscaled data sets. The RMSEs of the BCCAv2 data mainly reflect their systematic dry biases (figures not shown). Again, the different PF calculation methods based on the AMS and PDS show quite consistent results, regardless of the method used.

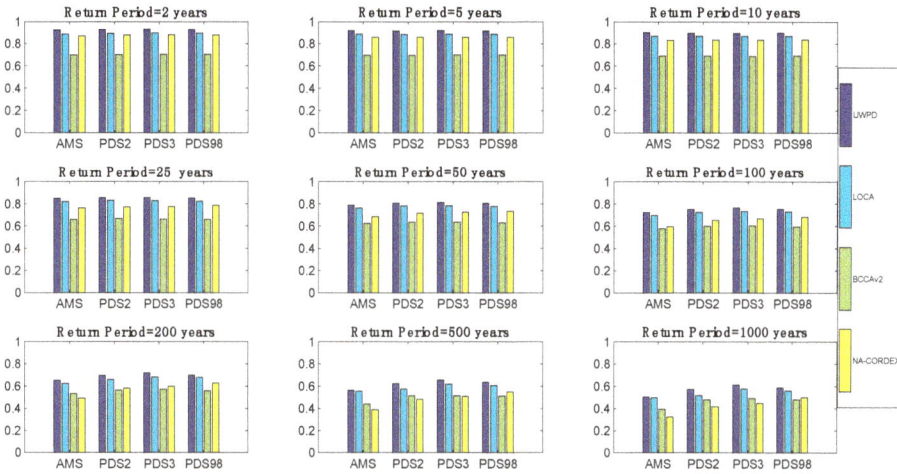

Figure 6. PF correlations between OB and downscaled data for different return periods. PF estimate correlation skills of the four downscaled data sets for different return periods: UWPD (dark blue), LOCA (light blue), BCCAv2 (light green), and NA-CORDEX (yellow).

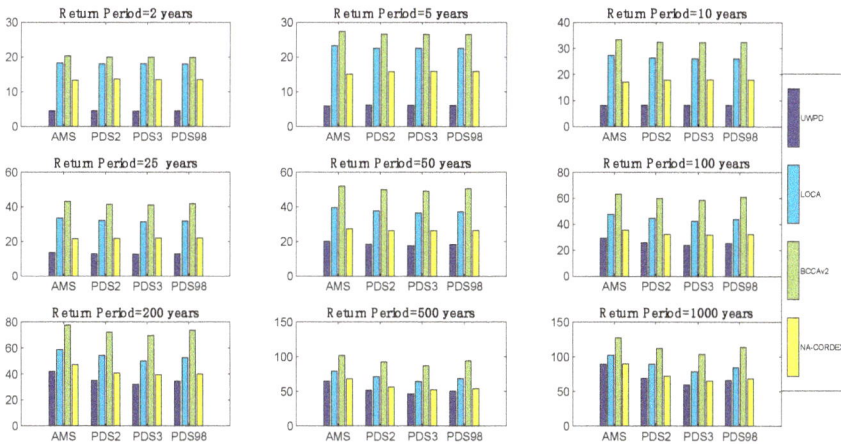

Figure 7. RMSE of PF estimates for downscaled data for different return periods. Similar to Figure 6 but for RMSE.

4. Projected PF Estimates

Based on our results in Section 3 in which the UWPD data were shown to capture major features of the observed PF estimates the best, we selected this data set for our primary analysis of future projections. Since the LOCA data set has been widely used by government reports and other publications [2], we also conducted parallel calculations based on this data set as well. Comparisons of the corresponding UWPD and LOCA results helps to determine how significant the findings really are. Doing so also gives some insight into other studies where the LOCA data have been used [4]. Our comparison study for the historical period in Section 3 showed that the PF estimates calculated by the different methods of AMS and PDS are quite similar. Therefore, for simplicity, we only show the PF estimate results that are based on the AMS for the following sections.

The projected data cover the period of 2006–2100. Knowing that a trend appears in this 95-year period, we divided the whole period into two smaller time periods: 2006–2053 (representing the first half of the 21st century) and 2054–2100 (representing the second half). Within each time period, we assumed the PF estimates are stationary as we did for the historical period, although some trends may also be detectable within each period. Figures 8 and 9 show for the low emission (RCP4.5) and high emission (RCP8.5) scenarios, respectively, the relative PF difference ratio comparing the projected data with the historical period. The ratio is defined as (PF_future – PF_historical)/PF_historical. PF estimates for return periods of 25 years and 100 years are shown here as representative examples.

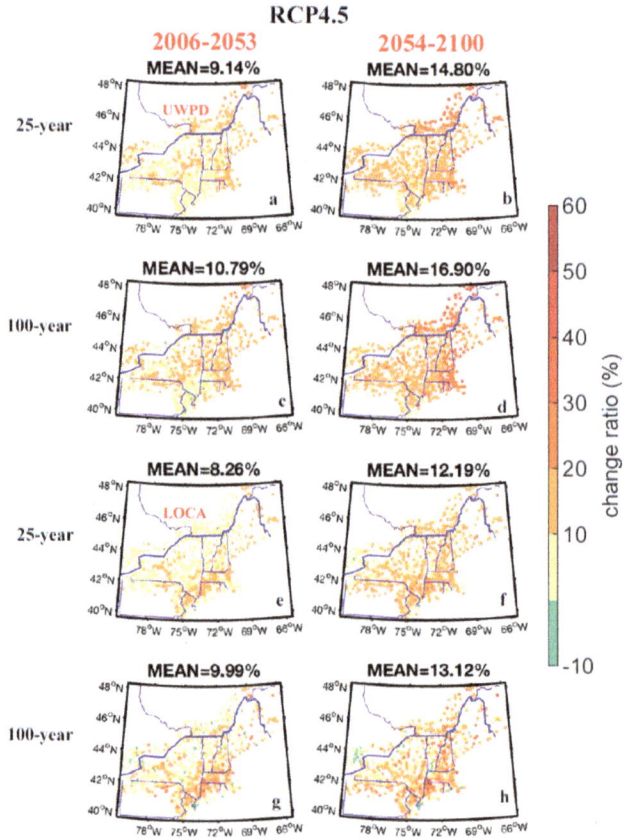

Figure 8. Relative PF difference ratios defined as (PF_future – PF_historical)/PF_historical for a 25-year return period (**a,b,e,f**) and a 100-year return period (**c,d,g,h**) under the RCP4.5 emission scenario.

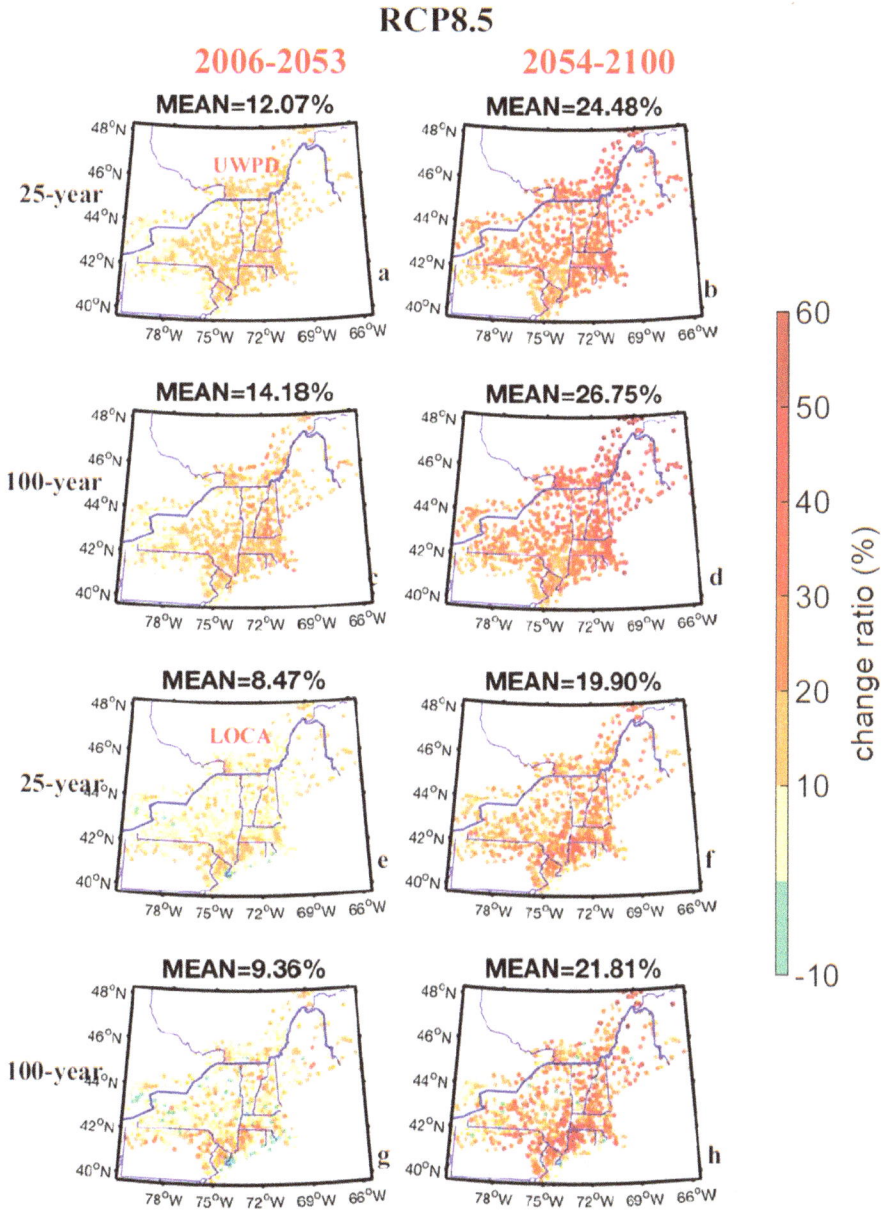

Figure 9. Relative PF difference ratios defined as (PF_future – PF_historical)/PF_historical for a 25-year return period (**a,b,e,f**) and a 100-year return period (**c,d,g,h**) under the RCP8.5 emission scenario.

Under the low emission scenario (RCP4.5) for the first half of the 21st century, the results based on the UWPD data show that, for a return period of 25 years (Figure 8a), PF estimates increase by a mean relative ratio of 0.09, which translates to a 9% increase. New Jersey and the southeast corner of New York experience the smallest change (0%–5%), while the area north of Maine experiences the largest change (20%–30%). Results for a return period of 100 years (Figure 8c) share quite similar spatial patterns with an average increase of 11%. Interestingly, several stations around New Jersey and the southeast corner of New York even show some negative values (−5%–0%), indicating a decrease of PF estimates. This decrease of PF estimates in some areas is also visible for the LOCA results for the return periods of 25 and 100 years (Figure 8e,g). In general, the results based on the LOCA data show relatively smaller increases with an average increase of 8% for a return period of 25 years and 10% for a return period of 100 years. The spatial distribution is also more scattered than for the UWPD data. PF estimates increase more dramatically by the second half of the 21st century. The UWPD data show an increase of 15% (Figure 8b) and 17% (Figure 8d) for return periods of 25 years and 100 years, respectively, while the LOCA data show an increase of 12% (Figure 8f) and 13% (Figure 8h).

Under the high emission scenario (RCP8.5), the magnitude of PF estimate changes is in general larger than for the low emission scenario. For the first half of the 21st century, the results based on the UWPD data show that PF estimates increase by a mean value of 12% for a return period of 25 years (Figure 9a) and 14% for a return period of 100 years (Figure 9c). Again, the LOCA data show smaller increases of 8% and 9% for return periods of 25 years and 100 years, respectively (Figure 9e,g). By the second half of the 21st century, however, the magnitudes increased quite dramatically, with both the UWPD and LOCA data sets showing average increases of over 20% (Figure 9b,d,f,h). For some stations around northern Maine, the UWPD data even show an increase of 50% for a return period of 100 years (Figure 9d). The LOCA data also have several stations with increases of over 50% (Figure 9h), but these stations are not as concentrated in one place as they are for the UWPD data. These location differences may be due to the ability of the models or downscaling schemes to simulate extratropical cyclones, which have been shown to be the main cause of extreme precipitation events in the Northeastern US [42]. The number of these cyclones is indeed projected to increase in the future [43]. Nevertheless, the PF estimate increases seem to be robust even though the distributions are not spatially uniform nor the same across the different downscaling data sets. Another way to look into the PF estimate changes is through the so-called exceedance probability, which will be discussed in the next section.

5. Changes to Exceedance Probabilities

The exceedance probability is the probability that a given rainfall total accumulated over a given duration (daily in our paper) will be exceeded in any one year. Compared with the related concept of a return period, it may sometimes be better to use the exceedance probability instead since many misconceptions are often associated with return periods. For example, there may be a misunderstanding that once an event happens, a certain number of years (the return period) must pass for an event of that magnitude to happen again. However, extreme events can happen in any year, regardless of how much time has elapsed since the previous one, an idea more easily conceptualized by the exceedance probability. These misconceptions associated with return periods are already discussed in depth [44]. Nonetheless, the term "return period" is still widely used in the literature, which is one reason why we employed it in our analyses in the previous sections. To complement these analyses, in this section, we examined future extreme precipitation events through the concept of exceedance probability. Specifically, we examined what the exceedance probabilities of the historical PF estimates will be in the future.

The probability density function for a GEV distribution associated with an AMS is given by:

$$P_{pdf} = \frac{1}{\sigma} t(x)^{1+\xi} e^{-t(x)}, \tag{2}$$

where $t(x) = \left(1 + \xi\left(\frac{x-\mu}{\sigma}\right)\right)^{-\frac{1}{\xi}}$, if $\xi \neq 0$ and $t(x) = e^{-\frac{(x-\mu)}{\sigma}}$ if $\xi = 0$. $\mu \in \mathbb{R}$ is the location parameter; $\sigma > 0$ is the scale parameter; and $\xi \in \mathbb{R}$ is the shape parameter.

Note that sometimes the shape parameter, ξ, is defined differently, often having opposite signs in different sources [40,45,46]. The cumulative density function (CDF) can be written as:

$$P_{cdf} = e^{-t(x)}. \tag{3}$$

The relevant exceedance probability, p, for a certain value, x, that is, the probability that a given year's maximum daily precipitation is larger than x, is given by:

$$p = 1 - P_{cdf} = 1 - e^{-t(x)}. \tag{4}$$

Since the CDF is invertible, the quantile function for the GEV distribution has an explicit expression. For any given exceedance probability, p, the related quantile is given by:

$$x = P_{cdf}^{-1}(1 - p). \tag{5}$$

For a given return period, T, in years, the exceedance probability for the PF associated with that period is $\frac{1}{T}$ (remembering that only one value per year is used in an AMS). Therefore, the PF can be calculated by plugging $\frac{1}{T}$ into Equation (5):

$$PF = P_{cdf}^{-1}\left(1 - \frac{1}{T}\right). \tag{6}$$

For any given AMS data, the GEV distribution parameters, μ, σ, and ξ, can be estimated by using methods, such as the maximum likelihood method and the L-moment method [36,42]. Parameters estimated based on the historical data are labeled here (μ_{hist}, σ_{hist}, ξ_{hist}, PF_{hist}) and parameters estimated based on the projected data are labeled (μ_{future}, σ_{future}, ξ_{future}). Likewise, PF estimates calculated based on the historical and projected data are labeled PF_{hist} and PF_{future}, respectively. To examine how the exceedance probabilities of the historical PF estimates will change in the future, we can compare the following:

$$P_{hist} = 1 - P_{cdf}\left(\mu_{hist}, \sigma_{hist}, \xi_{hist}, PF_{hist}\right) = \frac{1}{T}, \tag{7}$$

$$P_{future} = 1 - P_{cdf}\left(\mu_{future}, \sigma_{future}, \xi_{future}, PF_{hist}\right). \tag{8}$$

In Equation (7), we calculate the historical exceedance probability, P_{hist}, using distribution parameters estimated from the historical data and the PF_{hist} corresponding to a given time period T. In Equation (8), we instead calculate the future exceedance probability, P_{future}, using distribution parameters estimated from the projected future data, while still using the historical PF_{hist}. Since μ, σ, and ξ are different in each equation, we can expect different exceedance probabilities from each for the same PF estimates (or the same return periods).

Figure 10 shows averaged exceedance probabilities for the Northeastern US calculated from Equations (7) and (8) using the UWPD and LOCA data under the RCP4.5 and RCP8.5 emission scenarios. As expected, the projected exceedance probabilities (black and green lines) are all larger than the corresponding historical exceedance probabilities (red lines). For example, the historical 2-year event (total daily rainfall ≈60 mm) has an exceedance probability of 0.5 for one year. Under the RCP4.5 scenario, the UWPD data (Figure 10a) show that this amount will have an exceedance probability of ≈0.59 for the first half of the 21st century and 0.65 for the second half of the 21st century. These changes are more dramatic for larger events. For a historical 25-year event (total daily rainfall ≈107 mm), the exceedance probability changes from ≈0.04 to ≈0.07 for the first half of the 21st century and to 0.09 for the second half. For a historical 100-year event (total daily rainfall ≈138 mm), the exceedance probability changes from ≈0.01 to ≈0.03 for the first half of the 21st century and to ≈0.04 for the second half. The future exceedance probabilities are almost quadruple their historical values. This means a 100-year event will become a 35-year (1/0.0286 precisely, Table 2) event for the first half of the 21st century and a 27-year (1/0.037 precisely) event for the second half of the 21st century. Under the RCP8.5 scenario, the UWPD data (Figure 10b) show that the exceedance probability for a 100-year event may increase to 0.05 for the second half of 21st century, which is five times larger than its historical value; a 100-year event will become a 19-year event. The LOCA data (Figure 10c,d) in general tell a similar story, although the magnitudes are relatively smaller, probably related to the dry bias seen earlier in the data. Under the RCP8.5 scenario, the exceedance probability for the 100-year event (0.01) will increase to ≈0.04 for the second half of the 21st century, which is still quite a significant change. Different ensemble members indeed show different magnitudes (reflected by the short vertical lines in Figure 10 representing the ensemble spread), but overall, the signs of the changes are quite consistent, suggesting that the extreme events will become more frequent in the future.

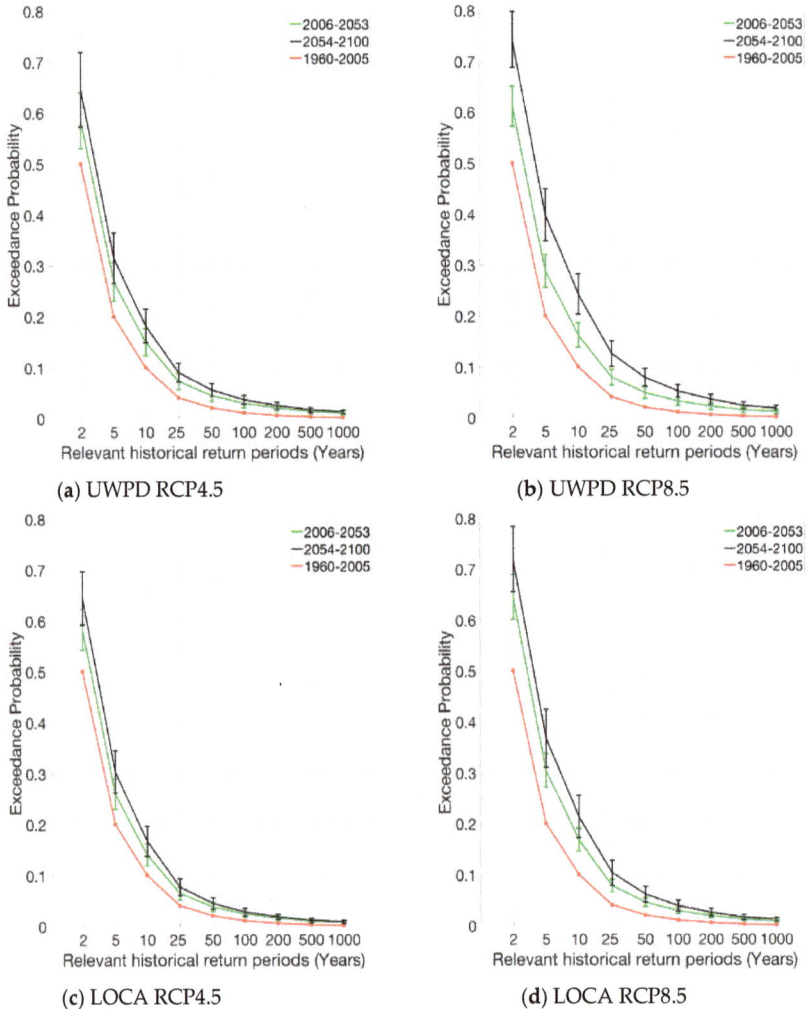

Figure 10. Northeastern US domain averaged exceedance probabilities (Equations (7) and (8)) for different relevant historical return periods (related to PF_{hist}) under different scenarios. Results for UWPD (upper; (**a**): RCP4.5; (**b**): RCP8.5) and LOCA (lower; (**c**): RCP4.5; (**d**): RCP8.5). The short vertical lines represent the uncertainties estimated by the ensemble spread. Red represents the results for the historical period (1960–2005), green for the first half of the 21st century (2006–2053), and black for the second half (2054–2100).

Table 2. Northeastern US domain averaged return periods with their uncertainties expressed as +/− one standard deviation of the ensemble under RCP4.5 and RCP8.5 scenarios.

Relevant Historical Return Periods (Years)		2	5	10	25	50	100	200	500	1000
Projected return periods (UWPD)	RCP4.5 2006–2053	1.7 [1.6–1.9]	3.7 [3.3–4.3]	6.7 [5.7–8.1]	13.8 [11.3–17.8]	22.7 [18.0–30.6]	34.9 [27.0–49.5]	50.9 [38.5–75.3]	77.2 [56.9–120.1]	100.4 [72.9–161.3]
	RCP4.5 2054–2100	1.5 [1.4–1.7]	3.2 [2.7–3.8]	5.5 [4.6–6.7]	11.0 [9.1–13.9]	17.8 [14.5–22.9]	27.2 [22.0–36.0]	39.5 [31.5–53.2]	59.9 [47.0–82.5]	77.8 [60.6–108.8]
	RCP8.5 2006–2053	1.6 [1.5–1.7]	3.5 [3.1–3.9]	6.2 [5.3–7.2]	12.6 [10.5–15.7]	20.6 [16.6–27.0]	31.6 [24.8–43.6]	45.9 [34.9–66.9]	69.4 [51.0–108.3]	90.0 [64.7–147.7]
	RCP8.5 2054–2100	1.3 [1.2–1.5]	2.5 [2.2–2.9]	4.1 [3.5–4.9]	7.9 [6.6–9.8]	12.5 [10.3–16.0]	19.0 [15.3–25.0]	27.4 [21.8–37.0]	41.4 [32.3–57.8]	53.9 [41.6–76.5]
Projected return periods (LOCA)	RCP4.5 2006–2053	1.7 [1.6–1.8]	3.9 [3.5–4.4]	7.1 [6.2–8.4]	15.6 [13.1–19.4]	26.7 [21.7–34.5]	42.7 [34.1–57.1]	64.5 [50.7–88.7]	102.0 [78.8–144.7]	136.2 [104.0–197.1]
	RCP4.5 2054–2100	1.5 [1.4–1.7]	3.3 [2.9–3.8]	5.9 [5.1–7.2]	12.9 [10.6–16.4]	22.0 [17.7–28.9]	35.2 [27.9–47.9]	53.3 [41.4–74.8]	84.1 [63.6–123.9]	112.0 [83.2–171.1]
	RCP8.5 2006–2053	1.7 [1.6–1.8]	3.7 [3.3–4.2]	6.9 [5.9–8.1]	15.1 [12.7–18.7]	26.0 [21.5–33.0]	42.0 [34.1–54.6]	63.7 [51.0–84.7]	101.1 [79.8–138.0]	135.2 [105.6–187.9]
	RCP8.5 2054–2100	1.4 [1.3–1.5]	2.7 [2.4–3.2]	4.7 [3.9–5.8]	9.6 [7.8–12.5]	16.1 [12.8–21.5]	25.4 [20.1–34.7]	38.2 [29.9–52.9]	60.0 [46.6–84.3]	79.8 [61.7–113.1]

6. Discussion and Conclusions

The observed global mean surface temperature has increased by about 1.0 °C (0.8–1.2 °C) above pre-industrial levels [47]. Associated with this temperature rise, the atmosphere's ability to hold water vapor has also increased, as described by the Clausius–Clapeyron relationship, implying a possible increase in extreme precipitation events [48]. Indeed, observational data show that extreme precipitation events in many parts of the world have increased in intensity dramatically over the past several decades, raising concerns about future changes. To address these concerns, many approaches based on observational and climate model data have been used. Despite significant uncertainties [49], climate models are used in numerous research publications to study projected climate changes and their consequences [2].

In contrast to the traditional approach of studying per-event precipitation intensities [50], in this study, based on observational and model projected data, we examined the point precipitation frequency (PF) estimates for the periods of 1960–2005, 2006–2053, and 2054–2100 in the Northeastern US, where the magnitudes of extreme precipitation events are projected to increase the most dramatically in the US [5]. Spatially, the Northeastern US is characterized by a highly diverse climate influenced by several geographic factors, such as the Atlantic Ocean to the east, the Great Lakes to the west, and the Appalachian Mountains to the south. The mountain ranges often block air flow, leading to local enhancement of precipitation through orographic lift [4]. To study the extreme precipitation events in such a complex region, we evaluated three commonly used downscaled data sets (LOCA, BCCAv2, and NA-CORDEX), along with a probabilistic downscaled data set developed specifically for this study (UWPD), on their ability to reproduce observed PF estimate features. This analysis was performed using four approaches, including one AMS-based and three PDS-based methods, but the variability in the projected PF estimates among these methods was not deemed significant. The data set with the highest historical accuracy statistics (UWPD) and one of the most commonly used data sets (LOCA) [2] were selected to analyze their projected PF estimates. Our results demonstrate that these PF estimates are likely to continue increasing, regardless of the emission scenario. The largest increases were in the northernmost part of the domain while the southern areas showed smaller increases. Under the RCP4.5 scenario, for the UWPD data set, the average projected increases to the 100-year event are 10.79% for the first half of the 21st century and 16.90% for the second half. The corresponding increases under RCP8.5 are 14.18% for the first half of the 21st century and 26.75% for the second half. This is generally consistent with the results of previous CMIP3 analyses of annual precipitation increases [4]. Our results also highlight how the associated exceedance probabilities of the historical PF estimates will change with projected climates. Notably, an event with a current exceedance probability of 0.01 (a 100-year event) may have an exceedance probability for the second half of the 21st century of ≈0.04 (a

27-year event) under the RCP4.5 scenario and ≈0.05 (a 19-year event) under RCP8.5, according to the UWPD data. It is thus vital to understand how these extreme precipitation events are projected to change so that the design decisions of today can serve to protect the socio-economic and environmental interests of the future.

Author Contributions: S.W. and M.M. designed the research framework. S.W. carried out the analysis. M.M. acted as the project leader. D.L. generated the UWPD downscaled data set and provided the description of this data. J.R.A. and K.G. participated in discussions and manuscript preparation.

Funding: The paper is supported by NOAA/UCAR Contract # SUBAWD000255.

Acknowledgments: We are grateful to Sanja Perica for her valuable comments and suggestions. We also thank Michael St Laurent for sharing the NOAA Atlas 14 observational data. Linda Mearns provided valuable discussions on the NA-CORDEX data; Katja Winger helped with storage issues on the NA-CORDEX data. David W. Pierce and Ben Livneh provided explanations about the LOCA data set, and Kenneth Nowak explained the BCCAv2 data set. Discussions with Brian Beucler during our online meeting were also quite helpful. Lisa Sheppard (Illinois State Water Survey) provided editorial assistance.

Conflicts of Interest: The authors declare no conflict of interest.

References

1. Milly, P.C.D.; Betancourt, J.; Falkenmark, M.; Hirsch, R.M.; Kundzewicz, Z.W.; Lettenmaier, D.P. Stationarity is dead: Whither water management? *Science* **2008**, *319*, 573–574. [CrossRef] [PubMed]
2. Easterling, D.R.; Kunkel, K.E.; Arnold, J.R.; Knutson, T.; LeGrande, A.N.; Leung, L.R.; Vose, R.S.; Waliser, D.E.; Wehner, M.F. Precipitation change in the United States. In *Climate Science Special Report: Fourth National Climate Assessment*; Wuebbles, D.J., Fahey, D.W., Hibbard, K.A., Dokken, D.J., Stewart, B.C., Maycock, T.K., Eds.; U.S. Global Change Research Program: Washington, DC, USA, 2017; Volume, I, pp. 207–230. [CrossRef]
3. IPCC. *Climate Change 2014: Synthesis Report. Contribution of Working Groups I, II and III to the Fifth Assessment Report of the Intergovernmental Panel on Climate Change*; Pachauri, R.K., Meyer, L.A., Eds.; IPCC: Geneva, Switzerland, 2014; 151p.
4. Kunkel, K.E.; Stevens, L.E.; Stevens, S.E.; Sun, L.Q.; Janssen, E.; Wuebbles, D.; Rennells, J.; Degaetano, A.; Dobson, J.G. *Regional climate trends and scenarios for the U.S. National Climate Assessment: Part. 1—Climate of the Northeast*; NOAA Technical report NESDIS 142-1; U.S. Department of Commerce: Washington, DC, USA, 2013.
5. Melillo, J.M.; Richmond, T.C.; Yohe, G.W. *Climate Change Impacts in the United States: The Third National Climate Assessment*; U.S. Government Printing Office: Washington, DC, USA, 2014; p. 841.
6. Hoerling, M.J.; Eischeid, J.; Perlwitz, X.W.; Quan, K.W.; Cheng, L. Characterizing recent trends in U.S. heavy precipitation. *J. Clim.* **2016**, *29*, 2313–2332. [CrossRef]
7. Huang, H.P.; Winter, J.M.; Osterberg, E.C.; Horton, R.M.; Beckage, B. Total and extreme precipitation changes over the Northeastern United States. *J. Hydr.* **2017**, *18*, 1783–1798. [CrossRef]
8. Maidment, D.R. *Handbook of Hydrology*; McGraw-Hill, Inc.: New York, NY, USA, 1993.
9. Hosking, J.R.M.; Wallis, J.R. *Regional Frequency Analysis: An Approach Based on L-Moments*; Cambridge University Press: Cambridge, UK, 1997.
10. Markus, M.; Wuebbles, D.J.; Liang, X.-Z.; Hayhoe, K.; Kristovich, D.A.R. Diagnostic analysis of future climate scenarios applied to urban flooding in the Chicago metropolitan area. *Clim. Chang.* **2012**, *111*, 879–902. [CrossRef]
11. Katz, R.W. Statistical methods for nonstationary extremes. In *Extremes in a Changing Climate: Detection, Analysis and Uncertainty*; Kouchak, A., Easterling, D., Hsu, K., Schubert, S., Sorooshian, S., Eds.; Springer: Dordrecht, The Netherlands, 2013; pp. 15–37.
12. Cheng, L.; AghaKouchak, A.; Gilleland, E.; Katz, R.W. Non-stationary Extreme Value Analysis in a Changing Climate. *Clim. Chang.* **2014**. [CrossRef]
13. Janssen, E.; Wuebbles, D.J.; Kunkel, K.E.; Olsen, S.C.; Goodman, A. Observational- and model-based trends and projections of extreme precipitation over the contiguous United States. *Earth's Future* **2014**, *2*, 99–113. [CrossRef]

14. Markus, M.; Angel, J.; Byard, G.; McConkey, S.; Zhang, C.; Cai, X.; Notaro, M.; Ashfaq, M. Communicating the Impacts of Projected Climate Change on Heavy Rainfall Using a Weighted Ensemble Approach. *J. Hydrol. Eng.* **2018**, *23*, 04018004. [CrossRef]

15. Randall, D.A.; Wood, R.A.; Bony, S.; Colman, R.; Fichefet, T.; Fyfe, J.; Kattsov, V.; Pitman, A.; Shukla, J.; Srinivasan, J.; et al. Cilmate Models and Their Evaluation. In *Climate Change 2007: The Physical Science Basis*; Contribution of Working Group I to the Fourth Assessment Report of the Intergovernmental Panel on Climate Change; Solomon, S., Qin, D., Manning, M., Chen, Z., Marquis, M., Averyt, K.B., Tignor, M., Miller, H.L., Eds.; Cambridge University Press: Cambridge, UK; New York, NY, USA, 2007.

16. Kunreuther, H.; Gupta, S.; Bosetti, V.; Cooke, R.; Dutt, V.; Ha-Duong, M.; Held, H.; Llanes-Regueiro, J.; Patt, A.; Shittu, E.; et al. Integrated Risk and Uncertainty Assessment of Climate Change Response Policies. In *Climate Change 2014: Mitigation of Climate Change*; Contribution of Working Group III to the Fifth Assessment Report of the Intergovernmental Panel on Climate Change; Edenhofer, O., Pichs-Madruga, R., Sokona, Y., Farahani, E., Kadner, S., Seyboth, K., Adler, A., Baum, I., Brunner, S., Eickemeier, P., et al., Eds.; Cambridge University Press: Cambridge, UK; New York, NY, USA, 2014.

17. Notaro, M.; Lorenz, D.J.; Hoving, C.; Schummer, M. Twenty-first-century projections of snowfall and winter severity across central-eastern North America. *J. Clim.* **2014**, *27*, 6526–6550. [CrossRef]

18. Vavrus, S.J.; Notaro, M.; Lorenz, D.J. Interpreting climate model projections of extreme weather events. *Weather Clim. Extrem.* **2015**, *10*, 10–28. [CrossRef]

19. Ashfaq, M.; Ghosh, S.; Kao, S.C.; Bowling, L.C.; Mote, P.; Touma, D.; Rauscher, S.A.; Diffenbaugh, N.S. Near-term acceleration of hydroclimatic change in the western U.S. *J. Geophys. Res.* **2013**, *118*, 10676–10693. [CrossRef]

20. Ashfaq, M.; Bowling, L.C.; Cherkauer, K.; Pal, J.S.; Diffenbaugh, S.N. Influence of climate model biases and daily-scale temperature and precipitation events on hydrological impacts assessment: A case study of the United States. *J. Geophys. Res. Atmos.* **2010**, *115*, D14116. [CrossRef]

21. Whetton, P.; Macadam, I.; Bathols, J.; O'Grady, J. Assessment of the use of current climate patterns to evaluate regional enhanced greenhouse response patterns of climate models. *Geophys. Res. Lett.* **2007**, *34*, L14701. [CrossRef]

22. Knutti, R. The end of model democracy? *Clim. Chang.* **2010**, *102*, 395–404. [CrossRef]

23. Perica, S.; Pavlovic, S.; Laurent, M.S.; Trypaluk, C.; Unruh, D.; Martin, D.; Wilhite, O. *Precipitation-Frequency Atlas of the United States. Version 3.0: Northeastern Statates*; National Weather Service: Silver Spring, MD, USA, 2019.

24. Stacy, E.W. A generalization of the gamma distribution. *Ann. Math. Stat.* **1962**, *33*, 1187–1192. [CrossRef]

25. Kirchmeier, M.C.; Lorenz, D.J.; Vimont, D.J. Statistical downscaling of daily wind speed variations. *J. Appl. Meteor. Climatol.* **2014**, *53*, 660–675. [CrossRef]

26. Daly, C.; Gibson, W.P.; Taylor, G.H.; Doggett, M.K.; Smith, J.I. Observer bias in daily precipitation measurements at United States cooperative network stations. *Bull. Amer. Meteor. Soc.* **2007**, *88*, 899–912. [CrossRef]

27. Kalnay, E.; Kanamitsu, M.; Kistler, R.; Collins, W.; Deaven, D.; Gandin, L.; Iredell, M.; Saha, S.; White, G.; Woollen, J.; et al. The NCEP/NCAR 40-year reanalysis project. *Bull. Am. Meteor. Soc.* **1996**, *77*, 437–471. [CrossRef]

28. Kirchmeier-Young, M.C.; Lorenz, D.J.; Vimont, D.J. Extreme Event Verification for Probabilistic Downscaling. *J. Appl. Meteor. Climatol.* **2016**, *55*, 2411–2430. [CrossRef]

29. Downscaled Climate Projections introduction. Available online: https://djlorenz.github.io/downscaling2/main.html (accessed on 17 June 2019).

30. Pierce, D.W.; Cayan, D.R.; Thrasher, B.L. Statistical downscaling using Localized Constructed Analogs (LOCA). *J. Hydrometeorol.* **2014**, *15*, 2558–2585. [CrossRef]

31. U.S. Global Change Research Program (USGCRP). *Impacts, Risks, and Adaptation in the United States: Fourth National Climate Assessment*; U.S. Global Change Research Program: Washington, DC, USA, 2018; Volume II, p. 1515. [CrossRef]

32. LOCA Statistical Downscaling (Localized Constructed Analogs). Available online: http://loca.ucsd.edu/ (accessed on 17 June 2019).

33. Maraun, D.; Wettterball, F.; Ireson, A.M.; Chandler, R.E.; Kendon, E.J.; Widmann, M.; Brienen, S.; Rust, H.W.; Sauter, T.; Theme, M.; et al. Precipitation downscaling under climate change: Recent developments to bridge the gap between dynamical models and the end user. *Rev. Geophys.* **2010**, *48*, RG3003. [CrossRef]

34. Downscaled CMIP3 and CMIP5 Climate and Hydrology Projections. Available online: https://gdo-dcp.ucllnl.org/downscaled_cmip_projections/dcpInterface.html (accessed on 17 June 2019).

35. Mearns, L.; McGinnis, S.; Korytina, D.; Scinocca, J.; Kharin, S.; Jiao, Y.; Qian, M.; Lazare, M.; Biner, S.; Winger, K.; et al. The NA-CORDEX dataset, version 1.0. In *NCAR Climate Data Gateway*; The National Center for Atmospheric Research: Boulder, CO, USA, 2017. [CrossRef]

36. The North American CORDEX Program—Regional Climate Change Scenario Data and Guidance for North America, for Use in Impacts, Decision-Making, and Climate Science. Available online: https://na-cordex.org/ (accessed on 17 June 2019).

37. Miller, J.F.; Frederick, R.H.; Tracey, R.J. *Precipitation Frequency Atlas of the Western United States*; National Technical Information Service: Springfield, VA, USA, 1973; Volume 11.

38. Perica, S.; Pavlovic, S.; Laurent, M.S.; Trypaluk, C.; Unruh, D.; Wilhite, O. *NOAA Atlas 14 Precipitation-Frequency Atlas of the United States*; Version 2.0; National Weather Service: Silver Spring, MD, USA, 2018; Volume 11.

39. Madsen, H.; Pearson, C.P.; Rosbjerg, D. Comparison of annual maximum series and partial duration series methods for modeling extreme hydrologic events regional modeling. *Water Resour. Res.* **1997**, *33*, 759–769. [CrossRef]

40. Coles, S. *An Introduction to Statistical Modelling of Extreme Values*; Springer: Berlin, Germany, 2001.

41. Langbein, W.B. Annual floods and the partial-duration flood series. Transactions. *Am. Geophys. Union* **1949**, *30*, 879–881. [CrossRef]

42. Kunkel, K.E.; Easterling, D.R.; Kristovich, D.A.R.; Gleason, B.; Stoecker, L.; Smith, R. Meteorological causes of the secular variations in observed extreme precipitation events for the conterminous United States. *J. Hydrometeor.* **2012**, *13*, 1131–1141. [CrossRef]

43. Ulbrich, U.; Leckebusch, G.C.; Grieger, J.; Schuster, M.; Akperov, M.; Bardin, M.Y.; Feng, Y.; Gulev, S.; Inatsu, M.; Keay, K.; et al. Are Greenhouse Gas Signals of Northern Hemisphere winter extra-tropical cyclone activity dependent on the identification and tracking algorithm? *Meteorol. Z.* **2013**, *22*, 61–68. [CrossRef]

44. Serinaldi, F. Dismissing return periods! *Stoch. Environ. Res. Risk Assess.* **2015**, *29*, 1179–1189. [CrossRef]

45. McFadden, D. Modeling the Choice of Residential Location. *Transp. Res. Record.* **1978**, *673*, 72–77.

46. Hosking, J.R.M. Moments or L moments? An example comparing two measures of distributional shape. *Am. Stat.* **1992**, *46*, 186–189. [CrossRef]

47. IPCC. Summary for Policymakers. In *Global Warming of 1.5 °C*; An IPCC Special Report on the impacts of global warming of 1.5 °C above pre-industrial levels and related global greenhouse gas emission pathways, in the context of strengthening the global response to the threat of climate change, sustainable development, and efforts to eradicate poverty; Masson-Delmotte, V., Zhai, P., Pörtner, H.O., Roberts, D., Skea, J., Shukla, P.R., Pirani, A., Moufouma-Okia, W., Péan, C., Pidcock, R., et al., Eds.; World Meteorological Organization: Geneva, Switzerland, 2018; 32p.

48. Kunkel, K.E.; Karl, T.R.; Brooks, H.; Kossin, J.; Lawrimore, J.H.; Arndt, D.; Bosart, L.; Changnon, D.; Cutter, S.L.; Doesken, N.; et al. Monitoring and understanding trends in extreme storms: State of knowledge. *Bull. Amer. Meteor. Soc.* **2013**, *94*, 499–514. [CrossRef]

49. Wang, C.Z.; Zhang, L.P.; Lee, S.K.; Wu, L.X.; Mechoso, C.R. A global perspective on CMIP5 climate model biases. *Nat. Clim. Chang.* **2014**, *4*, 201–205. [CrossRef]

50. Sharifi, E.; Steinacher, R.; Saghafian, B. Multi time-scale evaluation of high-resolution satellite-based precipitation products over northeast of Austria. *Atmos. Res.* **2018**, *206*, 46–63. [CrossRef]

MDPI

St. Alban-Anlage 66

4052 Basel

Switzerland

Tel. +41 61 683 77 34

Fax +41 61 302 89 18

www.mdpi.com

Water Editorial Office

E-mail: water@mdpi.com

www.mdpi.com/journal/water

www.ingramcontent.com/pod-product-compliance
Lightning Source LLC
Chambersburg PA
CBHW051900210326
41597CB00033B/5967